This volume makes an important contribution to the emerging critical literature on travel and the body. Its essays trace historical developments in the ways that travel is experienced as an embodied practice, and offer a thought-provoking range of perspectives and approaches. *Traveling Bodies* shows how productive interdisciplinary conversations in the field of travel can be, with the study of travel writing, enriched by attention to other forms of creative and cultural practice that explore bodies on the move. It is a book that couples broad and wide-ranging discussion of key ideas with detailed and thoughtful analytical work with particular examples and case studies, and will no doubt stimulate new journeys of exploration.

Zoë Kinsley, Associate Professor in English Literature,
Liverpool Hope University, UK

Up until today, matters of the body have not gained as much attention by international travel studies as they deserve. Therefore, the interdisciplinary essay collection *Traveling Bodies* can be considered a groundbreaking contribution to the field: It presents a wide range of new insights concerning (European, North American, and Japanese) travel literature and culture from the Age of Enlightenment to the present and offers innovative theoretical perspectives which will prove extremely productive for future work on the subject area. A highly recommended, almost indispensable read for scholars and students in travel studies around the world!

Stefan Hermes, Senior Lecturer in German Literature,
University of Duisburg-Essen, Germany

Traveling Bodies

Traveling Bodies: Interdisciplinary Perspectives on Traveling as an Embodied Practice explores the central role the body has in and for traveling and thus complements and expands upon existing research in travel studies with new perspectives on and insights in the entanglement of bodies and traveling. The case studies assembled in this volume discuss a variety of traveling practices, experiences, and media with chapters featuring Asian, American, and European historical and contemporary perspectives. Truly interdisciplinary in its approach, the volume identifies and examines diverse literary, historical, and cultural texts, contexts, and modes in which traveling and the body intersect, including 'classic' travelogues, (new) media (e.g., film and digital travel apps), surf culture, and travel-inspired tattoos. The contributions offer various avenues for further research not only for scholars working with body theory and travel (writing) but also for anyone interested in the intersections of literature, culture, media, and embodied practices of traveling.

Nicole Maruo-Schröder is Professor of Cultural Studies at the University of Koblenz, Germany. Her research focuses on nineteenth-century American literature, material (food) culture, travel writing, intersectionality, and visual culture. Publications include co-edited collections on *Literature and Consumption in Nineteenth-Century America* (2014), *Space, Place, and Narrative* (2016), and *Issues in Contemporary Young Adult Dystopian Fiction* (2018) as well as a monograph on *Spatial Concepts in Contemporary American Literature* (2006). A current book project focuses on literature and consumption.

Sarah Schäfer-Althaus is Lecturer in Anglophone Literature and Culture at the University of Koblenz, Germany. Her research centers on women, gender, sexuality studies, and medical humanities. She is the author of *The Gendered Body: Female Sanctity, Gender Hybridity and the Body in Women's Hagiography* (2016) and co-editor of *Transient Bodies in Anglophone Literature and Culture* (2020) and *Medicine and Mobility in Nineteenth-Century British Literature, History, and Culture* (2023).

Uta Schaffers is Professor of German Literature and Didactics at the University of Koblenz, Germany. Her main research areas include travel writing (various articles and the co-edited volume *(Off) the Beaten Track? Normierungen und Kanonisierungen des Reisens*; 2018) with special focus on Japan (*Konstruktionen der Fremde. Erfahren, verschriftlicht und erlesen am Beispiel Japan*; 2006) and the Swiss travel writer Annemarie Schwarzenbach (see the editions of Schwarzenbach's works), traveling bodies, and East-Asia in literature, as well as economics and literature.

Routledge Research in Travel Writing
Edited by Peter Hulme, University of Essex and Tim Youngs,
Nottingham Trent University

The Desertmakers
Travel, War and the State in Latin America
Javier Uriarte

Time and Temporalities in European Travel Writing
Paula Henrikson and Christina Kullberg

Revisiting Italy
British Women Travel Writers and the Risorgimento (1844–61)
Rebecca Butler

Travel, Travel Writing, and British Political Economy
"Instructions for Travellers," circa 1750-1850
Brian P. Cooper

The Arabian Desert in English Travel Writing Since 1950
A Barren Legacy?
Jenny Walker

Ethics of Description
The Anthropological Dispositif and French Modern Travel Writing
Matt Reeck

Travel Writing and Re-Enactment
Echotourism
Lucas Tromly

Traveling Bodies
Interdisciplinary Perspectives on Traveling as an Embodied Practice
*Edited by Nicole Maruo-Schröder, Sarah Schäfer-Althaus, and
Uta Schaffers*

For more information about this series, please visit: https://www.routledge
.com/Routledge-Research-in-Travel-Writing/book-series/RRTW

Traveling Bodies

Interdisciplinary Perspectives on Traveling as an Embodied Practice

Edited by
Nicole Maruo-Schröder,
Sarah Schäfer-Althaus, and
Uta Schaffers

Routledge
Taylor & Francis Group

NEW YORK AND LONDON

First published 2024
by Routledge
605 Third Avenue, New York, NY 10158

and by Routledge
4 Park Square, Milton Park, Abingdon, Oxon, OX14 4RN

Routledge is an imprint of the Taylor & Francis Group, an informa business

ISBN: 9781032360911 (hbk)
ISBN: 9781032364117 (pbk)
ISBN: 9781003331803 (ebk)

DOI: 10.4324/9781003331803

Typeset in Sabon
by Deanta Global Publishing Services, Chennai, India

Contents

Figures

List of Contributors

Anne Barjolin-Smith holds a doctorate in American Studies and is an independent researcher living in Florida, USA. As a specialist and participant in lifestyle sports, she travels the world to explore regional snowboarding and surfing cultures, confronting them with their American versions. Her current research focuses on the notion of coloniality in lifestyle sports and surf music.

Sofie Decock is Associate Professor at the Department of Translation, Interpreting and Communication at Ghent University, Belgium. She conducts research in the fields of professional communication and German Studies with a focus on (persuasive) discourse, (intercultural, digital) interactions, and gender.

Karly Etz is Postdoctoral Associate in Art History at the Rochester Institute of Technology, USA. Her research investigates the recent inclusion of tattooing within contemporary art spaces and the medium's ability to offer unique points of entry into an array of pressing art historical concerns in the twenty-first century.

Sonja Klein is Assistant Professor (Akademische Rätin auf Zeit) for German Studies at Heinrich Heine University Düsseldorf, Germany. Her research areas include art and the body, cultural memory, literary anthropology, poetry, and plant studies. Her second book (Habilitation) focuses on body symbolism in the writings of Johann Wolfgang von Goethe.

Nicole Maruo-Schröder is Professor of Cultural Studies at the University of Koblenz, Germany. Her research focuses on nineteenth-century American literature, material (food) culture, travel writing, intersectionality, and visual culture. Publications include co-edited collections on *Literature and Consumption in Nineteenth-Century America* (2014), *Space, Place, and Narrative* (2016), and *Issues in Contemporary Young Adult Dystopian Fiction* (2018) as well as a monograph on *Spatial Concepts in Contemporary American Literature* (2006). A current book project focuses on literature and consumption.

Michael Meyer is Professor of Anglophone Literatures and TESOL at the University of Koblenz, Germany. Among his publications are articles and edited collections on Romantic literature, modern and postmodern British literature, colonial and postcolonial literature, and film. He wrote monographs on British autobiography, British poetry, an *Introduction to English and American Literatures* (4th ed. 2011), and co-wrote *Teaching English* (2nd ed. 2022). His current research interest focuses on travel writing, postcolonial literature, and visual media.

Andreas Niehaus is Professor of Japanese Studies at Ghent University, Belgium. His research focuses on early modern and modern body cultures and sports in Japan. Publications include the co-edited volumes *Feeding Japan: Cultures and Politics of Food Identities* (2017) and *Sport, Memory, and Nationhood in Japan: Remembering the Glory Days* (2013). He is also co-founder and speaker of the Centre for Research on Body Cultures in Motion (BOCULT).

Sarah Schäfer-Althaus is Lecturer in Anglophone Literature and Culture at the University of Koblenz, Germany. Her research centers on women, gender, sexuality studies, and medical humanities. She is the author of *The Gendered Body: Female Sanctity, Gender Hybridity and the Body in Women's Hagiography* (2016) and co-editor of *Transient Bodies in Anglophone Literature and Culture* (2020) and *Medicine and Mobility in Nineteenth-Century British Literature, History, and Culture* (2023).

Uta Schaffers is Professor of German Literature and Didactics at the University of Koblenz, Germany. Her main research areas include travel writing (various articles and the co-edited volume *(Off) the Beaten Track? Normierungen und Kanonisierungen des Reisens*; 2018) with special focus on Japan (*Konstruktionen der Fremde. Erfahren, verschriftlicht und erlesen am Beispiel Japan*; 2006) and the Swiss travel writer Annemarie Schwarzenbach (see the editions of Schwarzenbach's works), traveling bodies, and East-Asia in literature, as well as economics and literature.

Mira Shah earned her doctorate in the study of literature from Bern University, Switzerland, with a thesis on the poetics and politics of emotionality in primatology before exploring a biocultural geography of the Pacific islands with a project on 'Entangled Island Times' at Goethe-University Frankfurt, Germany. Currently, she is concerned with the history, culture, literature, and presence of the Stone Age and the ways we colonize the past to make it fit our present's needs.

Anne von Petersdorff-Campen is an independent scholar and filmmaker. She received her PhD in German Studies and Digital Humanities from Michigan State University, USA, and works at the intersection of creative practice and research. She currently lives in Berlin, Germany, where she works as a producer for international documentary films.

Nora Winsky is Research Associate at the Institute of Environmental Social Sciences and Geography at the University of Freiburg, Germany. In 2022, she completed her doctoral degree in the field of tourism studies with a thesis on tourist practices and related representations in the city of Freiburg and the Black Forest. Currently, she is working in the research fields of tourism geography as well as geographic aging studies.

Elizabeth Zold is Professor of English at Winona State University, USA. Her research focuses on eighteenth-century women's travel literature, and her scholarship has been published in such journals as *Studies in Travel Writing, Pedagogy,* and *ABO: Interactive Journal for Women in the Arts, 1640–1830.*

Acknowledgments

The initial idea for this volume was born amid lively and fruitful discussions at the conference "Traveling Bodies" held back in the summer of 2018 at the University of Koblenz-Landau. Discussing our ideas about traveling and its intersections with the body, little did we know that in the near future, a pandemic would bring the world to a sudden standstill. Our own traveling bodies were forced to remain immobile, our exploration and discovery of new spaces and places limited to armchair traveling in the comfort of our homes. This volume is one of the results of these armchair explorations. We are especially obliged to Tim Youngs, who, during one of the conference's coffee breaks, encouraged us to turn our ideas into a proposal for an edited collection in the *Routledge Research in Travel Writing* series. We would like to thank him and co-editor Peter Hulme for the opportunity to publish our book as part of the series.

This project would not have been possible without our contributors, whose inspiring insights make up this collection and whom we would like to thank for their time, commitment, and patience. Some of us have kept working together in exploring the role that the body and corporeality have for the experience of traveling as well as for writing about it. This collaboration has turned from an informal into an institutionalized research network on "Traveling Bodies/Reisende Körper", funded by the German Research Foundation (DFG), and we are looking forward to working in this fascinating field in the years to come.

Many thanks go to Gisela Anheier (*in memoriam*) and our excellent (former) student assistants Maximilian Bähr, Michelle Bebbon, Sarah Rörig, and Lana Shelemey for helping us organize the initial conference. We are also grateful for the generous funding for it provided by the 'Forschungsfond' of the University of Koblenz-Landau. Moreover, we would like to thank our colleagues in the Faculty of Arts and Humanities in Koblenz for their support, interest, and many inspiring discussions. Special thanks go to Jennifer Abbot and Anita Bhatt at Routledge for all their work in producing this book. Finally, we are indebted to our editorial assistant Marie Kluge for her invaluable help in preparing the manuscript.

In the course of the project, some of us had to travel onwards, whereas others have crossed our paths and even joined our travels – to all those traveling bodies this work is dedicated.

March 2023   Nicole Maruo-Schröder
                                                      Sarah Schäfer-Althaus
                                                          Uta Schaffers

1 Traveling Bodies

An Introduction

Nicole Maruo-Schröder, Sarah Schäfer-Althaus, and Uta Schaffers

Our backs as well as the leather seats were soaked with perspiration. When we stood for a while in the great wind we felt chilled till we were dry, the evaporation of water in this very dry air being almost as rapid as that of pet- rol. Nevertheless our progress was more comfortable than my overcrowded bus journey of two years ago. After days of rattle-trapping along that same road, my back had rubbed so much against a plank that I found my clothes worn through when I reached the garage at Shahrud; at that time my back was 'steeled', for a lumbar pain obliged me to travel with a metallic corset. That inconvenience added to the heat of a Persian summer made things a little difficult; and camp- ing for days together by the road-side where breakdowns had stranded us, it was impossible to undress. I tried to cheer myself, remembering what Irene had told me about her father Lord Curzon who journeyed to the Pamirs encased in a similar contrivance, for he suffered from a chronic sciatica.[1]

This quote from Ella Maillart's travelogue *The Cruel Way. Switzerland to Afghanistan in a Ford, 1939* presents two female travelers[2] on a tour in the region of Khorasan (Iran) between Shahrud and Mashhad – a region Maillart characterized earlier as "a dull part of the world, a succession of immense waterless basins separated by hills of grey gravel or by the rocky ribs of a skeletal earth" (loc. 1704–24). While driving through a dusty, windy "oven-like desert" (loc. 1724), the climate is nearly unbear- able: "Before noon, the heat was already lording it over our world. Every detail of the country quivered as if at the point of boiling" (loc. 1724). The description of the inconveniences the travelers face serves as a start- ing point for a reminiscence of another journey which was even more difficult and ends with a short note on a traveler who faced similar or worse challenges on his journey to the Pamir Mountains. What connects all three scenes is the fact that they bring the traveler's body into focus, be it because it is affected by climate or physical impairment, be it because of the climate adding to the difficulties that arise from a severe health condi- tion and a pain-ridden body.

Travelers are always traveling bodies, bodies on the move. Our body as a "dynamic locus of human thought, action, and language"[3] is the medium

DOI: 10.4324/9781003331803-1

of our travels, and our mobility, our perceptions, and experiences strongly depend on its condition and its functioning. The limitations of the body, for example, in terms of the experience of space, can be overcome by technological means, as the range and the pace of the spatial experience are expanded when traveling by car or bus, by ship or plane; and thanks to body enhancements, it is possible to go on but also to continue a journey even with "lumbar pain" and "chronic sciatica". The experiences and memories of the 'travel writer'[4] in Maillart's travelogue are presented and formed with rhetorical strategies and figures that 'cross-fade' means of the traveler's body, mobility, and the machine. In the first scene, sweat penetrates the seats of the car ("Our backs as well as the leather seats were soaked with perspiration") and in the 'lording heat' and dry air, body fluid evaporates as fast as petrol – it almost seems as if the heat in the region of Khorasan has similar qualities as the heat of an internal combustion machine.[5] Additional analogies and strategies of 'blending' evoke even more images of the transgression of boundaries between body and machine, and even the transformation of one into the other: on another journey that in the first quote is presented as a retrospection, the back of the travel writer was 'steeled with a metallic corset' while sitting on a bus, enclosing the travelers' body in two metallic containers. These circumstances make the journey possible in the first place but at the same time difficult (or nearly unbearable). Like all technological means of transportation and mobility, a car or a bus disconnects the traveler from direct bodily contact with and physical experience of the new environment: captured in a 'metallic container', the traveler has to slow down, get out, and immerse his or her (lived) body in this new environment in order to truly experience it with all senses.

As Maillart's impressions of her journey from Geneva to Kabul so aptly show, traveling is a thoroughly embodied practice as we encounter and explore the world through our sense perceptions, which are not passive impressions but active processes in our interaction with the world.[6] Traveling and, by extension, travel writing is therefore as much about the new, the unfamiliar, and the strange, as it is about the familiar, about ourselves, and what we already know. Bodies play a central role in both aspects.[7] While this seems obvious to point out, the body and its significant role in and for traveling and travel writing is one of the areas that have remained largely unexplored in the study of travel literature, a field that by now has become an established topic in academic research.[8] Although the (re-)presentation of the bodies of the so-called others has been discussed in travel writing, particularly from a postcolonial perspective, the traveler's own body and her sense perceptions have played a much lesser role.[9] Based on notions of the body as a 'lived body' and traveling as an embodied practice in which bodily experience is part and parcel of how the world 'appears' to us, the present volume, therefore, focuses on various intersections between traveling and the body in travel texts.

The transformation of the travel experience into a text is even more mediated than the actual experience of the traveler, a transformation that is affected by aspects as different as the time lag (between travel and writing) and the manifold conventions and prefigurations attached to the genre of the travel account.[10] The staging of the traveler's own bodily experience – particularly sense perceptions – as well as the encounter with the other's body is a significant part of the (re-)construction of the travel experience in such texts. Such descriptions can serve to enable the reader to 'feel' – see, hear, smell, taste, and touch – what the traveler claims to have experienced at a certain time and in a certain place. Truthful or not, these experiences are part of the rhetoric of travel writing,[11] which comprises a multitude of different aspects from the writer's attempt to 'visualize' – bring to life – scenes and sights encountered during the journey to the adherence to genre conventions. Sense perceptions, therefore, play an essential part in the ways in which the unfamiliar has been staged and constructed as the traveler's 'other', often inferior, yet also slightly dangerous.[12] While we, as embodied beings, cannot encounter the unfamiliar on neutral ground, with a "naked eye"[13] so to speak, it is important to explore the complex ways in which bodies – our selves – are part of the construction of other places, other cultures, and other people; in short, what it means to conceptualize traveling as an embodied experience, and how and with which effects the body, sensory perceptions, and bodily experiences are presented and staged in travel writing.

Furthermore, traveling means mobility. It means going 'elsewhere', even if this is only imagined, as is the case with armchair traveling. Hence, travelers are figures on the move, in mind and body, which means that they experience themselves, their own bodies but also other bodies in encounters on their journeys, in unfamiliar surroundings and unfamiliar ways. Just like Maillart on her journey to Iran, the experience of traveling can make us experience our bodies in new ways. The demands of climatic conditions, unusual foodways, or exertions, such as Maillart's sore back, can lead to unexpected bodily reactions: we can feel uncomfortable and even fall ill when faced with the challenges of travel. New sensual experience can, of course, also impact us in a positive way, for example, the relaxation we experience simply because we leave our (tedious) everyday lives behind for a while or the pleasure we feel when looking at an awe-inspiring landscape:

> And then, O reward! the parched desert was cloven by the intense indigo of a river flowing towards us. Half blind from the glare of a torrid noon, we rushed to the shade of a massive old bridge, jumped

out of our sticky clothes and, poised on a rock, looked at the miraculous blueness that was to enfold us. But, damp with perspiration, we shivered in the great wind though the temperature of the air must have been above 115 degrees. And quickly plunged. Up to our necks in the river, our bodies light as feathers, all our fatigue wiped away, we were in a kind of paradise, soft waters gliding smoothly around our limbs, eyes filled with the azure that surrounded us while hurried ripples raced each other eagerly, joyfully.

(Maillart, loc. 2058–66)

Maillart's experience of the swim in the river as a stay in paradise after escaping the hell-like heat of "the parched desert" is a case in point. Thus, these 'new' bodily experiences can lead to a fresh look at what we are used to, a (self-)reflection that can range from appreciation to renegotiation and dismissal of both, what we are used to and what is new to us. We can even become 'estranged' from our own bodies since our body schemes and images as well as familiar body practices and techniques may no longer work or fit, challenged by the need to adapt to the unfamiliar.[14]

Traveling, however, does not just provide us with a fresh view of ourselves. It also entails, and indeed most often its point is, the experience of the 'new' (including also medial prefigurations) or the unusual in the form of nature, landscapes, (material) cultures, and, most importantly, people. This encounter of the new and unfamiliar is in complex ways entangled with who we are and what we already know. The norms, values, and ideologies that we are used to, our entire 'way of living', are embodied in a variety of ways, practices, and rituals as has been variously shown,[15] and in this sense, they accompany us on our journeys. Perception itself is "not to be construed as the passive mirroring or representation of the outer world by a subject",[16] but as "an activity of exploring the environment".[17] Alva Noë points out that "[p]erception is not something that happens to us, or in us. It is something we do. [...] [W]e *enact* our perceptual experience; we act it out".[18] What is more, it is a "knowledgeable activity",[19] an interaction with the world that is influenced by what we know (including attitudes, norms, values, and narratives) and what we expect. In that sense, the way we look at others is shaped by schemes, patterns of reference, and attitudes that are culture-based, collective, as well as individual,[20] and which we use to evaluate and even judge these 'others':

I was interested in a group of women and children who had come on pilgrimage: I had never seen such people before. They all had something thick and ugly about them, almost simian, with fuzzy hair and enormous mouths. I remembered that according to the legend, demoniac races used to live on the shores of the Caspian, far away from the plateaux where the Aryans dwelt. A little girl with frizzy hair and thick lips looked as if she were snarling at me: it robbed us of the courage to ask these people who they were.

(Maillart, loc. 1597–98)

When Maillart writes about her encounter with the pilgrims, she emphasizes their absolute difference: the travel writer "had never seen such people". Moreover, she 'sees' that difference as something negative and deviant as to her the pilgrims appear "ugly", ape-like, and hairy, a perception based on long-standing tropes within travel writing that categorize the other as inferior, sub-human, even animal-like. Although she claims that she "remembered" the legend of "demoniac races" as a result of her encounter with the pilgrims, one could argue that it could also be the other way around: the legend, as a prefiguration, might have triggered the evaluation and classification of the observed pilgrims. The look of the girl – perceived and judged by the travel writer "as if she were snarling at me" – can be read as a subversion of the traveler's controlling gaze and is reason enough for her not to establish any further contact. As a result, the pilgrims remain a legend, not least for the readers.

Particularly in the West, vision plays a central role in exploring 'new' worlds.[21] Coming from an ocularcentric culture, looking at something or someone suggests distance and control as well as truth and objectivity. Looking provides us with a sense of superiority (on our part) as well as the feeling of availability (on the part of the person looked at). Moreover, when rendering our travel experience for others, vision becomes a guarantor of truth and knowledge – being an eyewitness is evidence for the authenticity of the traveler's experience, her 'having been there', which is why photography has played such an outstanding role within travel and its various modes of documentation. That vision is just one dimension of the travel experience becomes particularly clear when reading travel accounts of blind people.[22] As Margaret Topping asserts, "the presumption that value is guaranteed by vision and visibility is disrupted by travelers who are 'deprived' of sight".[23] This dominance of sight over the other senses does not only become obvious in its central role within the documentation of travel; it is also the reason for the neglect of the significance of the other senses within the study of traveling and travel accounts.[24]

Similar to our vision, the smelling and tasting of food, for instance, is heavily affected by sociocultural preconceptions of what is tasty and edible, and new foodways or smellscapes can have an important influence on how we judge our travel experience and the people we encounter. Moreover, unfamiliar experiences can trigger – by way of smell or taste – memories of past experiences, which in turn affect the present moment and our perception. In this sense, the old and the new, the known and the strange, the self and the other are not polar opposites but rather points of interconnection and entanglement in the complex web of our embodied experience of the world. Particularly when sight is restricted or absent, the other senses take over and compensate – in the experience of travel as much as in its documentation. When vision is limited, more attention seems to be paid to what the other senses can offer. However, that the other senses are (or should be) by no means a mere substitute becomes clear when we think about how senses and places can be connected in terms of food ('the taste of France') or sound ('the sound of Cuba') and even smell (as in smell-walker Kate McLean's "smellmaps").[25] These senses, however, are in no way more 'neutral' than vision, but just as much caught up in sociocultural preconceptions, which therefore become part of our perception of other people and other places. While vision helps us to 'keep our distance', elevates us, and gives us a certain feeling of control, this is not necessarily the case with other sense perceptions: "Water from the wells is often too brackish: to train your system to cope with it you are advised to eat raw onions with buttermilk. And the millions of flies are far too sticky and affectionate. But once you have risen above these small inconveniences, Iran is yours" (Maillart, loc. 1078).

Touch, for instance, requires us to be very close, a proximity to the new and unfamiliar[26] that is not always appreciated as Maillart's comment on the "sticky and affectionate" flies documents. In this sense, touch can also be more immediate, more intensive than looking, it can even be invasive. For Maillart, for example, her bath in the river and in particular the feel of the water is intensely refreshing for her whole body, whereas the idea of lips and tongue touching the "brackish" waters from the wells and the thought of ingesting and swallowing it is repellent. Moreover, touching something gives us more information on shape, texture, and surface than vision does and can thus be more personal and intimate. Similarly, smell and taste seem more immediate and bring us closer to whatever and whoever we encounter: both penetrate the body in ways different from looking or touching. Taste, in particular, requires us to ingest something, thus opening our body to the unfamiliar, making it potentially 'vulnerable'; in more extreme cases, it can even lead to feelings of disgust and nausea (and rather violent bodily reactions). Food and its taste, not surprisingly, are therefore of central concern for the traveler, expressing the necessity to adapt bodily needs, habits, and preferred taste to the available food and eating habits. It is also very much an aspect of the evaluation of the new and unfamiliar – particularly, it seems, when the evaluation turns out to be negative. Maillart's comment on the brackish water and the 'strange' remedy of eating onions together with buttermilk can serve as an example here, one that might even be called one of the more harmless judgments regarding other people's eating habits. Read as laconic, her remark on

rising "above such slight inconveniences" underlines the traveler's qualities of coping, stamina, and fearlessness in the face of new and potentially
dangerous places and people.[27] Read as ironic, however, her comment suggests the very opposite of what she seems to say, framing such eating habits
as repulsive and so offensive as to make an exploration, an experience of
the country almost impossible.

1.1 Traveling Bodies: The Contributions

As our discussion here has shown, any analysis of discourses about the
body in travel texts has to acknowledge the fact that the ways in which
other bodies and bodily practices are perceived and experienced are inevitably and inextricably intertwined with the traveler's and, respectively, the
travel writer's own bodily perceptions and experiences. The present collection is an attempt to open up ideas for this exploration. It collects a variety
of different disciplines and approaches, all of which take the body as the
focal point for their analysis of traveling (and related practices) and travel
writing. The collection is divided into four parts that highlight different
aspects of the intersection of bodies and traveling.

The contributions in the first part focus on the *Body as Concept and
Metaphor* within travel writing from the eighteenth to the nineteenth century. In her contribution "The Scientist-Traveler and the Woman-as-Land:
Sexual Topographies in *A New Description of Merryland* (1741)", Sarah
Schäfer-Althaus looks at the genre of somatopic erotica, in which particularly the female body was described through cartographic and topographic
metaphors. In this genre, new scientific discoveries and medical discourses
intersect with erotic images and imaginations of the (female) body, producing, as Schäfer-Althaus argues, satires of eighteenth-century travel literature, more particularly the Grand Tour, whose concepts of "male elite
education" and "pleasure and improvement"[28] are echoed. Bodies emerge,
in this type of literature, as both traveling (male) subjects and traveled to
(female) objects, offering us a unique, highly gendered portrayal of the
corporeality of traveling. Sonja Klein's "From Facts to Physicality: Body
Concepts in German Travel Writing Around 1800" shifts the focus to a
different genre and a different time. Concentrating on German travel writing around 1800, the contribution traces the development of the genre's
preference for scientific data, statistics, and detailed observation to a more
individual embodied experience of traveling. Klein discusses this transition
from an 'outside' world of the traveled regions to the 'inside' world of the
(male) traveler in Johann Wolfgang von Goethe's works – especially in
his *Letters from Switzerland. Part the First* (1808) – and their emphasis
on the subjective physicality of the embodied traveler experiencing both
the landscapes and his own body. In her "Motherhood and the Embodied

Traveler in Wollstonecraft's *Letters Written during a Short Residence in Sweden, Norway, and Denmark"* (1796), Elizabeth Zold argues that Mary Wollstonecraft uses her subject position as a mother traveling with her daughter to demonstrate the important connection between mind and body to increase the authenticity of travel writing. Wollstonecraft draws attention to the erasure of the embodied observer, demonstrating the ways in which the thinking and feeling self is needed to 'authentically' represent the fullness of the travel experience. Zold shows how the author utilizes her daughter's presence as a locus from which to explore the relationship between emotion and reason, and how acknowledging her embodiment enhances the truth-value of the narrative.

Other Bodies, the second part of this collection, highlights the concept of otherness and how it can be an essential part of travel experience and travel writing. In "Beasts on Board: Traveling Animals and Pacific Voyages in the First Two Ages of Exploration", Mira Shah examines the important role that animal bodies played in the conquest and exploration of the Pacific world and its documentation in travel accounts. In addition to their 'mundane' roles as food or means of travel, animal bodies also function as gifts and trading goods, and thus as a medium for the communication and negotiation of social norms and values in scenes of intercultural contact. Shah highlights these diverse roles by discussing the various conflicts that could arise when answering such simple questions as to which animal bodies were edible or which could serve as a gift. Analyzing the complexity of the ways in which such 'Animal Others' were perceived, Shah shows how they were used to express and also establish (colonial) power structures, on board the ships as much as in the intercultural encounters of the European travelers and their so-called others. Michael Meyer's contribution "Mary Wollstonecraft and the Body of her Letters, or: The Traveler Lost and Found in Scandinavia" provides us again with an analysis of Wollstonecraft's journey to Scandinavia, focusing, however, on the neglected private *Letters to Imlay* (1798) and their difference to the published travelogue from 1796. The private letters reveal what the published account does not cover, namely the very beginning and ending of her journey, in which the traveler's alienation at the outset of and the return from her travel – the threshold between home and abroad – is focused on. Using Bernhard Waldenfels's account of the embodied process of experiencing and understanding encounters with alien individuals and cultures, Meyer turns to this frame of Wollstonecraft's journey, showing how the traveler – ironically – profits from the foreign societies she encounters in reconfiguring her impaired and alienated self as superior to the locals. Accordingly, Wollstonecraft's empowerment is only possible because she complements her embodied alienation with the subjection of the other to her preconceived notions. Nicole Maruo-Schröder takes a look at travel accounts by women writers who traveled the American frontier during the first half of the nineteenth century in "'The Most Dirtiest Children': Spectacles of Otherness on the American Frontier". Rather than focusing on the familiar views of the landscape typical for the genre, her contribution concentrates on how bodies, and more particularly the bodies of the so-called 'others', are portrayed as sights. Looking at travel narratives by Eliza Farnham, Caroline Kirkland, Susanna Moodie, and Susan Shelby Magoffin, she argues that the travel writers stage what one could call 'spectacles of otherness', a narrative display of the bodies of others that seeks to impress upon readers the dangers that freedom can have for bodies that need disciplining. Focusing on class-based perception of otherness, Maruo-Schröder analyzes how such 'deviant' bodies threaten to 'contaminate' the frontier, sometimes even the writers themselves, a contamination that is especially expressed by the sensual experience (vision, smell, touch) and the bodily response of the travelers encountering 'the other'.

The third part focuses on *Crossing Borders: The Body and Its Liminal Zones*. In his contribution "'My Condition Gets Worse Day by Day': Controlling Traveling Bodies on the Move in Edo-Period Japan", Andreas Niehaus analyzes exemplary passages in travel writings of Japanese travelers in the seventeenth and eighteenth century that deal with bodies losing, trying to maintain, and regaining control. The loss of control is not only a result of the limitations of the body that constantly forces itself *as body* on the attention of the traveler but is also the result of political and social powers and ideologies (ranging from material barriers to social categorizations according to gender and class) during the Edo-period. It also exposes the travelers to situations where they constantly experience transgressions of social, emotional, and bodily liminal zones. From the beginning of the eighteenth century onward, bodies and bodily experience turned from mere metaphorical markers in travelogues to a text element that wanted to transmit a (somatic) reality, as Niehaus argues. Hence, the chapter focuses on the reciprocity between the experience of the traveling body and the way this corporeal experience is given textual form in diaries and poems that are part of travel texts. In the following contribution, the focus remains on Japan, but it zooms in on Westerners traveling in Japan in the late nineteenth and early twentieth century. Based on the reflection of the ineluctable necessity and corporeality of eating, the chapter "'The "Food Question" Is Said to Be the Most Important One for All Travelers': Eating in Travel Writing" by Uta Schaffers explores some of the intricate connections between traveling (bodies), food, eating, and the perception of the Other. While traveling, food as a "liminal substance"[29] can be experienced as a source of joy and/or of (self-)observation and irritation, and as such it is often a central topic in travel writing. Considering that the "oral cavity forms the contact zone between the interior world of the body and the exterior world of objects",[30] the (imagined or actual) incorporation of foodstuff is directed and experienced by all senses and can evoke strong feelings. As Schaffers shows, these bodily 'responses' to food might be perceived as being natural but are, in fact, deeply interwoven with, for example, cultural, ethical, religious, or national discourses and are used to defame and denigrate a culture or nation while at the same time underlining one's 'own' culture or nation as superior. In "Going Undercover? Female Bodies and Clothes under Scrutiny in Travel Literature", Sofie Decock examines the social semiotics of bodily practices and clothing performances in (the depiction of) intercultural encounters in Maillart's and Annemarie Schwarzenbach's travel writing on Afghanistan in the 1940s. As clothes are physically touching, (partly) covering, and protecting human bodies, they constitute a liminal zone between bodies, the environment, and the others, and play an important part in constructing, regulating, and managing our bodily practices, our public physical appearance, and the ways our bodily appearance is perceived and judged by others. Decock depicts the limitations and the transgression of traditional gender norms by analyzing the 'clothing management' of the two travelers in intercultural encounters and discusses various instances of 'othering' in descriptions of local Afghan women, concentrating on (the absence of) veiling practices and the body postures associated with them.

The fourth part of this volume, *Mobility, Perception, Experience*, collects contributions that look more concretely at the tensions between mobility and experience, and the perception of dis- and emplacement that emerge when bodies travel. Such a focus includes matters of the performance of bodies just as much as inscriptions of these travels onto bodies. In her ethnographic study "Surfing Wanderlust: Surf-Tripping Bodies as Cultural Bearers", Anne Barjolin-Smith analyzes surf trips as a fundamental element of the surfing experience and the surfer's traveling body as a political figure whose performance and representation are underpinned by cultural, ideological, historical, and geographical issues. The observed Floridian surfers' bodies carry an American mythology from their 'own' land to a 'foreign' one, the Maldives, and their practices mirror the complex subjectivities and singularities of their own surfing community. The chapter examines ways in which American surfers' wanderlust does not just showcase 'Americanness' (the quality of being American) but also enables 'Americanity' (understood as American domination founded on the idea of progress). In "Traveling Bodies in Film: Embodied Encounters and Negotiating Selves", Anne von Petersdorff-Campen takes a very different look at how bodies travel and what kind of traces are left in their encounters with others. Drawing on phenomenological approaches to film and feminist film studies, she analyzes filmic strategies that highlight the physical and social dynamics of embodied encounters of women travelers, more precisely how these are mediated and translated to an audience. Focusing on selected scenes and techniques in Ruth Beckermann's *Eine flüchtige Reise nach dem Orient* (1999) and Helke Misselwitz's *Tango Traum* (1985), von Petersdorff-Campen shows how filmic strategies such as 'haptic visuality' or 'embodied cinematography' are used in the negotiation of the filmmakers' subjectivity, creating a more fluid, ambiguous representation of their travel experience. Finally, referring to her own filmic work, the author connects the practical example of documenting the traveling body more closely to her theoretical insights into the role that the traveling body plays in the filmmaking process itself. In "Tattooed Cartographies and the Displaced Body in an Age of Political Conflict", Karly Etz conceptualizes tattoos as an itinerant medium grounded in lived experience, which allow wearers to visually map their travel through space and authenticate their experiences of displacement on those same traveling bodies. Etz examines the work of three artists, Qin Ga, Wafaa Bilal, and Douglas Gordon, who build on the tattoo's historical relationship to bodily mobility and systems of power to formulate critical geographies. The chosen works vividly demonstrate the ways in which tattooed cartographies speak to disrupted notions of homeland and shifting conceptual boundaries by virtue of their location on the body's liminal zone: skin. Each of the artworks examined within this chapter serves as a touchstone for discussing the thematic relevance of tattooed cartographies in relation to the experiences of forcibly displaced subjects, namely their relationship to issues of mobility, marginality, nationalism, liminality, borderlines, (in-)visibility, unity, and division among others. Nora Winsky's article "Strolling through the City on a Self-Guided Tour: Embodied Engagement with the Urban Space", finally, turns to the role digital technologies can play in the bodily experience of traveling, more specifically to the intertwining of digital and physical worlds. Taking a self-guided tour offered by *Freiburg Living History* as her example, Winsky explores the interactions between traveling bodies, urban settings, and mobile media from a phenomenological point of view. In her analysis, the technical features of the digital city tour, ranging from audio sequences about living conditions in the Middle Ages to visualizations of the architectural changes over time, help to stimulate the traveler's "sensory-inscribed" body,[31] enabling an experience in which material and immaterial – digital – elements of the city tour merge to provide the (bodily) experience of a time travel to the historical city of Freiburg at the same time as the present-day city is explored.

Putting together this collection during the time of the COVID-19 crisis has brought the importance of traveling and mobility, the (lived) body, and even various forms of perception and mediation into sharp relief. As the world came to a standstill – or so it felt to many of us – forms of traveling and the experience it comes with have fundamentally changed.

We were virtually – digitally – still mobile and 'traveling', yet our bodies were immobilized at the same time, forced to remain in one spot. While this is somewhat reminiscent of the so-called 'Kopfreise', the armchair traveling made possible by travel literature, it is by far not as satisfying. Impacting our perceptions and experience in fundamental ways, the forced immobility made the role that our bodies play during traveling more than evident. To us, this has clarified once again the fact that perception and experience are embodied, stressing the importance of looking at travel and traveling practices (in travel writing and beyond) in relation to the body.

Notes

1 Ella Maillart, *The Cruel Way. Switzerland to Afghanistan in a Ford, 1939* [1947] (Chicago: University of Chicago Press, 2013). Kindle edition (at loc. 1739–45). Further references to this edition are included parenthetically in the text.

2 The Swiss travel authors Ella Maillart and Annemarie Schwarzenbach traveled together in a new Ford from Geneva to Afghanistan. They started in June 1939 and arrived three months later in Afghanistan. See also Sofie Decock's contribution in this volume.

3 Mark Johnson, "What Makes a Body?", *The Journal of Speculative Philosophy*, 22, no. 3 (2008): 159–69 (at p. 159).

4 With the term 'travel writer' [Reiseschreiber], we want to stress the conceptual difference between the traveler him- or herself – the empirical person traveling and experiencing – and the traveler ('I' or 'me' in the text) as a media figure that is part of the text and the textual representation. Compare Alfred Opitz, *Reiseschreiber. Variationen einer literarischen Figur der Moderne vom 18.–20. Jahrhundert* (Trier: WVT, 1997).

5 Especially for women, the car became a central means of independent transportation and mobility that helped to override the conventional forms of perception and the handed-down topoi used to describe a certain landscape all over again. It certainly allowed them to expand the radius of traveling and to reach the more *unbeaten tracks*, but it also deeply changed the travelers' 'view' and visual perception (as chauffeur or co-driver alike). Beth Ann Muellner points out that exploring areas by car "might also be seen as an instrument of colonial expansion that serves to reinforce colonial rule." "Roving Reporter, Travel Journalist, Storyteller. Annemarie Schwarzenbach (1908–1942)", in Christa Spreizer, ed., *Discovering Women's History. German-Speaking Journalists (1900–1950)* (Oxford: Peter Lang, 2014), pp. 147–71 (at p. 163)). See also Esme Coulbert and Tim Youngs, eds., "Travel Writing and the Automobile", *Studies in Travel Writing*, 17, no. 2 (2013); Anke Hertling, *Eroberung der Männerdomäne Automobil. Die Selbstfahrerinnen Ruth Landshoff-Yorck, Erika Mann und Annemarie Schwarzenbach* (Bielefeld: Aisthesis, 2013); and Sofie Decock and Uta Schaffers, "Intercultural Encounters in Travel Texts of Ella Maillart and Annemarie Schwarzenbach: Perspectives from Literary Studies and Linguistics", in Gonçalo Vilas-Boas and Maria de Fátima Outeirinho, eds., *Annemarie Schwarzenbach e a literatura de viagens na Europa dos anos 30* (Porto: Edições Afrontamento, 2016), pp. 15–49.

6 Alva Noë, *Action in Perception* (Cambridge, MA: MIT Press, 2006).
7 See Zbigniew Białas, *The Body Wall: Somatics of Travelling and Discursive Practices* (Frankfurt: Peter Lang, 2006), p. 12, who states that "the traveller's body – whether hotly denied or not, whether saved or mutilated – remains at the very core of the representational enterprise of travel writing."
8 Apart from Białas, *The Body Wall*, see, for example, Marguerite Helmers and Tilar J. Mazzeo, eds., *Journal of Narrative Theory. Travel and the Body*, 35, no. 3 (2005); Eitan Bar-Yosef, "The 'Deaf Traveller', the 'Blind Traveller', and Constructions of Disability in Nineteenth-Century Travel Writing", *Victorian Review*, 35, no. 2 (2009): 133–54; Charles Forsdick, "Travel and the Body. Corporeality, Speed and Technology", in Carl Thompson, ed., *The Routledge Companion to Travel Writing* (Abingdon: Routledge, 2016), pp. 68–77; Alasdair Pettinger and Tim Youngs, eds., *The Routledge Research Companion to Travel Writing* (Abingdon: Routledge, 2020), pp. 181–262; and Nicole Maruo-Schröder and Uta Schaffers, "Körper", in Hansjörg Bay et al., eds., *Handbuch Literatur und Reise* (Metzler; forthcoming).
9 Białas claims an "awareness of the somatics of the traveller himself – his agreement to and with his own person or lack thereof *vis-à-vis* other bodies, his 'sensibility' (to use the eighteenth-century nomenclature), his translation of the somatic into the semantic" (*The Body Wall*, p. 14). See also Maruo-Schröder and Schaffers, "Körper", n. pag.
10 Ansgar Nünning, "Zur mehrfachen Präfiguration/Prämediation der Wirklichkeitsdarstellung im Reisebericht. Grundzüge einer narratologischen Theorie, Typologie und Poetik der Reiseliteratur", in Marion Gymnich et al., eds., *Point of Arrival: Travels in Time, Space, and Self. Zielpunkte: Unterwegs in Zeit, Raum und Selbst* (Tübingen: Francke, 2008), pp. 11–32.
11 See also Białas, *The Body Wall*, p. 21: "[T]he travelling somatic body becomes in itself a rather intricate instrument, not only articulating and receiving certain impulses but also, and more importantly, transporting specific meanings and creating persistent 'scapes' that enter travel discourse."
12 The description of unfamiliar terrain, of endeavors into *terra incognita*, as well as encounters with the other and their bodies as frightening and sometimes even threatening, has a long history. As Mary Baine Campbell notes with regard to travel writing between 400 and 1600: "The farther one got from Home, the temperate, reasonable mean, the more outlandish, *unheimlich*, became the bodies and manners of men" (*The Witness and the Other World: Exotic European Travel Writing, 400–1600* (Ithaca and London: Cornell University Press, 1988), p. 65).
13 Giorgia Alù and Sarah Patricia Hill, "The Travelling Eye: Reading the Visual in Travel Narratives", *Studies in Travel Writing*, 22, no. 1 (2018): 1–15 (at p. 2).
14 See, for example, Marcel Mauss, "Die Techniken des Körpers", in Wolf Lepenies and Henning Ritter, eds., *Soziologie und Anthropologie* 2 (München: Hanser, 1975), pp. 199–220; Shaun Gallagher and Jonathan Cole, "Body Schema and Body Image in a Deafferented Subject", *Journal of Mind and Behaviour*, 16 (1995): 369–90; and Shaun Gallagher, *How the Body Shapes the Mind* (Oxford: Oxford University Press, 2005).
15 For different disciplinary and theoretical angles, see, for example, Pierre Bourdieu, *Distinction. A Social Critique of the Judgement of Taste*, trans. Richard Nice (Cambridge, MA: Harvard University Press, 1984); Norbert Elias, *The Civilizing Process. Sociogenetic and Psychogenetic Investigations* [1939], trans. Edmund Jephcott, ed. Eric Dunning, Johan Goudsblom, and Stephen Mennell (Oxford, Malden: Blackwell, 2000); Lisa Heldke, *Exotic*

Appetites: Ruminations of a Food Adventurer (London: Routledge, 2003), p. 5; Joerg Fingerhut, Rebekka Hufendiek, and Markus Wild, eds., *Philosophie der Verkörperung* (Berlin: Suhrkamp 2013); and Matthias Jung, Michaela Bauks, and Andreas Ackermann, eds., *Dem Körper eingeschrieben. Verkörperung zwischen Leiberleben und kulturellem Sinn* (Wiesbaden: Springer, 2016).

16 Matthias Jung, *Science, Humanism, and Religion. The Quest for Orientation* (Cham: Palgrave Macmillan, 2019), p. 121.

17 Alva Noë, "Précis of Action in Perception: Philosophy and Phenomenological Research", *Philosophy and Phenomenological Research*, LXXVI, no. 3 (2008): 660–65 (at p. 664).

18 Noë, *Action*, p. 1 (emphasis in text).

19 Noë, *Action*, p. 3.

20 Kerstin Gernig, ed., *Fremde Körper. Zur Konstruktion des Anderen in europäischen Diskursen* (Berlin: Dahlem University Press, 2001).

21 See, for example, Mary Louise Pratt, *Imperial Eyes. Travel Writing and Transculturation* [1992] (New York: Routledge, 2008); and Alù and Hill, "The Travelling Eye", pp. 1–15.

22 Interestingly enough, as Bar-Yosef points out in "The 'Deaf Traveller'", the sense of vision still plays a significant role in travel accounts of the blind (pp. 144–46).

23 Margret Topping, "Seeing", in Pettinger and Youngs, eds., *Companion to Travel Writing*, pp. 193–207 (at p. 193).

24 Topping, "Seeing", p. 208.

25 Kate McLean, "Sensory Maps" (n.d.), https://sensorymaps.com/about/ [2 April 2022]; see also Clare Brant, "Smelling", in Pettinger and Youngs, eds., *Companion to Travel Writing*, pp. 249–61.

26 Sarah Jackson, "Touching", in Pettinger and Youngs, eds., *Companion to Travel Writing*, pp. 222–35.

27 See also Heidi Oberholtzer Lee, "Tasting", in Pettinger and Youngs, eds., *Companion to Travel Writing*, pp. 236–48.

28 Karen Harvey, *Reading Sex in the Eighteenth Century* (Cambridge: Cambridge University Press, 2004), p. 176.

29 David Bell and Gill Valentine, *Consuming Geographies: We Are Where We Eat* (London: Routledge, 1997), p. 44.

30 "Der Mundraum bildet mithin die Kontaktgrenze von Körperinnenwelt und objekthafter Körperaußenwelt" (Hartmut Böhme and Beate Slominski, "Einführung in die Mundhöhle", in Hartmut Böhme and Beate Slominski, eds., *Das Orale. Die Mundhöhle in Kulturgeschichte und Zahnmedizin* (Paderborn: Fink, 2013), pp. 11–31 (at p. 14), trans. Uta Schaffers).

31 Jason Farman, "Map Interfaces and the Production of Locative Media Space", in Rowan Wilken and Gerard Goggin, eds., *Locative Media* (New York: Routledge, 2014), pp. 83–93.

I

The Body as Concept and Metaphor

2 The Scientist-Traveler and the Woman-as-Land

Sexual Topographies in *A New Description of Merryland* (1741)

Sarah Schäfer-Althaus

2.1 Introduction

> Of all the Studies to which Men are drawn, either by Inclination or Interest, perhaps no one can pretend to such an agreeable Pleasure, as the DESCRIPTION OF COUNTRIES. By a Variety of Prospects, they feed us constantly with fresh Satisfactions; and the Objects they present are so chained together that a curious Reader has much ado to break off. This is the Advantage of that Subject in general.[1]

In the eighteenth century, at the time of British imperial expansionism, an increasing number of bodies were on the move, including travelers, explorers, scientists, colonizers, and imperialists who set out to expand both their knowledge of the world and the British territory, via close observation, systematization, mapmaking, documentation, and (violent) colonization of countries and continents, their topographies, dimensions, nature, customs, and inhabitants. These 'traveling bodies' and their purposes for traveling were indeed diverse, ranging from young gentlemen of the upper classes sent on a Grand Tour through Europe, a rite of passage "intended to prepare these young men to assume the leadership positions preordained for them at home",[2] to explorers and cartographers like James Cook, who navigated the *Endeavour* and *Resolution* on his three voyages to Australia and the South and North Pacific through uncharted territory.

No matter the journey's intention, these traveling bodies envisaged, and were anticipated by others, to actively participate in shaping this new age of discovery and reason with novel and valuable insights and to satisfy the Enlightenment's emphasis on empirical knowledge and experience.[3] To document their travels and observations, a vast amount of travel writing was produced and disseminated during the period, ranging from personal correspondence, journals, and travelogues to natural histories, botanical treatises, maps, and atlases. Particularly, "[j]ournalism and narrative travel accounts", as Mary Louise Pratt notes in *Imperial Eyes: Travel Writing and Transculturation* (1992), "were essential mediators between the scientific network and a larger European public".[4] Travel literature

DOI: 10.4324/9781003331803-3

enjoyed unprecedented popularity among the British, and many writers of the time, such as Lady Mary Wortley Montagu, Tobias Smollett, and Mary Wollstonecraft, produced publications once they had returned from their travels to Turkey, France, Italy, and Scandinavia.[5] By contrast, other novelists, like Aphra Behn, Daniel Defoe, and Jonathan Swift, reacted to the public's increasing interest in travel writing by publishing fictional accounts of travels to 'exotic' destinations.[6] In short, the eighteenth century "was the age of gold for travelers, both real and imaginary".[7]

Authors of erotic literature also fabricated *erotica curiosa* similar in structure and style to contemporary natural histories and travelogues. As Karen Harvey notes, "[i]n depicting bodies in motion, eighteenth-century erotica was shaped by geographical knowledge and its modes of expression",[8] producing texts in which imperial and erotic fantasy merged with travel literature conventions. One of the earliest book-length descriptions of the woman-as-land in erotica can be found in *Erōtopolis, the Present State of Betty-Land* (1684), a work attributed to Charles Cotton,[9] which depicts the female body as a country tilled and husbanded by men. In erotic culture, the female body was compared to a 'Merryland', "an imaginary place which promised the delights of sex and female bodies".[10] This metaphorical image of the woman-as-land is primarily associated with more liberal, secular contexts celebrating male sexuality and was employed most famously at the Beggar Benison's and Merryland of Anstruther, a gentleman's club founded in Scotland in 1732. In its records, "Happiness and Prosperity of [...] the Inhabitants of our Celebrated Territories of MERRYLAND" as well as the support of "Trade, Manufacture, and Agriculture in that delightful Colony" are listed as its central affairs. New members were given a diploma, which granted them "full powers and Privileges [sic] of INGRESS, EGRESS, AND REGRESS from and to and to and from all the Harbours, Creeks, Havens, and Commodious Inlets upon the Coasts of our said Extensive Territories".[11] Part of the eighteenth-century libertine tradition in literature, the concept of 'Merryland' thus indicated a "new literary attitude toward the erotic body", which Franz Reitinger terms "geo-pornography".[12]

Half a century after the publication of *Erōtopolis*, the Merryland theme culminated in the publication of a series of texts by Edmund Curll (1675–1747), a controversial publisher and bookseller of pirated editions and pornography.[13] The most popular text of said series, *A New Description of Merryland: Containing a Topographical, Geographical, and Natural History of That Country* (1741), allegedly written by Thomas Stretzer (of whom nothing is known), centers on a scientist-traveler's journey to, on, and into the female body ('Merryland').[14] The woman-as-land in *Merryland* is a 'somatopia',[15] passively enduring the experienced and heroic colonial man of science on his 'empirical' quest to observe, explore, systematize, and examine the supposedly foreign female 'Other'. It is a place "composed of bodies (female) *and* designed for bodily pleasure (male)".[16] *Merryland* reverses the land-as-woman metaphor common in colonial travel literature, which described and sexualized the conquered land as feminine. Moreover, by presenting the female form as a country, the erotic pamphlet illustrates that sex "was not simply regarded as an act – whether of reproduction, desire or violence; sex was a place for men to visit".[17]

Dedicated to Scottish physician George Cheyne (1671–1743) and pointed at "the curious and learned Body, the *Royal Society*" (*Merryland* 5), of whom Cheyne was a member, the satire is directed at a European, male, upper-class, and lettered audience.[18] The tribute to the learned society is significant as it implies that "its discoveries deserve the attention of the Royal Society, a joke made poignant by the interest the Society displayed in matters concerning generation and female sexuality".[19] Moreover, with its intersection of pleasure, knowledge, science, and empiricism, *Merryland* mirrors the ideological statutes behind the Grand Tour, anticipating that the formation of elite male identities should not only have an educative function but also include (sexual) adventures and experiences, albeit away from home.[20]

In the following, I seek to demonstrate that *Merryland* is more than just a Georgian "pornographic fantasy-travel book".[21] Profoundly allegorical, the pamphlet makes use of contemporary literary trends to address a kaleidoscope of the century's political, cultural, social, scientific, and gendered concerns. Located at the intersections of visual observation, physical exploration, and medical examination, the traveling I's gaze is simultaneously male, colonial, and medical when traveling to, on, and later in(to) the female body. Produced at a time when women became increasingly subjected to a controlling, authoritative, and professionalized male and medical gaze,[22] and, as I have argued elsewhere, "significant changes in thinking about men and women, sex and gender, and (reproductive) sexuality [...] challenged the traditional understanding of the human body",[23] *Merryland* echoes the complex discourses about women, femininity, and sexual difference, and reinforces the idea of men's physical and intellectual superiority prevalent in the eighteenth century.

2.2 From Exterior to Interior: Travelogue Conventions, Observation, and the Scientist-Traveler's Ordering Eye

By the eighteenth century, travel writing included textual features and conventions that, if followed, immediately identified the text as part of the travelogue genre and the author as a credible and trustworthy informant. *Merryland* masquerades as a travelogue by including a dedication to a patron and an address to the reader, providing a justification and truth claim for both journey and text, which is followed by the text proper, a geographic, topographic description of Merryland and its natural history.[24] In mimicking travel literature convention, erotic authors challenged "contemporary Enlightened geographical practices by producing their own distinctive brand of travel lie".[25]

Moreover, as Harvey notes, *Merryland* "was ostensibly written as a corrective to the work 'of modern Geographers', and to atlases, globes and modern histories which neglected to describe"[26] the female utopian landscape. Indeed, the alleged 'editor' of *Merryland* wonders in his address to the reader whether it is an increased "Fondness for Fairy-Tales, fabulous Stories, monstrous Fictions and Romances" that caused the subject's "long Neglect in *England*" at a time when "many of our Countrymen have been travelling abroad" and "much Diligence is daily made use of to procure any thing that is *new*" (vii). The travel writer likewise makes a nationalist claim by expressing his astonishment that particularly the "English Geographers take no Notice of it" (xi) and concludes that "their Silence has rendered this Work [...] the more necessary" (xiii), whose meticulous study will inform and at the same time entertain the reader (xiv).[27]

To establish its obscure 'author', Roger Pheuquewell, an Irishman, as a trustworthy and credible colonial man of science and "a potential Royal Society informant",[28] the text draws on complex rhetorical strategies. In the author's preface, the travel writer introduces himself as a man with "twenty Years Experience, and frequent Opportunities of acquainting [himself] with the Situation and Circumstances of MERRYLAND", who, having "finished [his] Inquiries into The Present State of that Country", is about to share his findings to provide future travelers with "a tolerable Idea" of Merryland (*Merryland* ix). Pheuquewell's implied advanced age and extensive travel experience express an almost natural seniority and authority on the topic, which is supported further by his travelogue's scientific design. His intention is not to "amuse Mankind with [...] uncertain Guesses" but "to examine what others have published" (x) and then "to compare them one with another, and with [his] own Observations, in order to sift out the Truth" (xi), mirroring empirical research methods and suggesting a rigorous scientific research design. Moreover, he promises to describe particularly those geographic curiosities and scientific mysteries of Merryland "well worthy the Consideration of that curious and learned Body, the *Royal Society*" (5) by pointing out the country's geographic curiosities and mysteries, such as its ever-changing longitude and latitude.[29]

Furthermore, Pheuquewell repeatedly insists on being an observer guided by controlled vision and reason: "I endeavoured to get the best Insight that was possible in every Thing relating to the State of MERRYLAND, observing with diligent Attention every thing the Country afforded that was remarkable" (*Merryland* 4). As Matthew Edney notes, "[o]f all the senses, sight was the most mechanistic, least sensuous, and hence was closest to the truth"[30] and thus to reason. Moreover, Pheuquewell's use of 'insight' in the quote is significant as it implies "sight with the 'eyes' of the mind"[31] and thus demonstrates the close connection between sight and visual perception and understanding and knowledge. Characterizing the traveler as an observer, as someone who "looks at the world in a controlled manner"[32] rather than a mere spectator, who is more interested in entertainment than detached objectivity, consequently perfects the illusion of Pheuquewell's credibility and authenticity and illustrates the century's emphasis on vision.[33]

In addition, the eighteenth century witnessed "a new orientation toward exploring and documenting continental interiors, in contrast with the maritime paradigm that had held center stage for three hundred years".[34] *Merryland* reflects this new scientific paradigm by presenting the reader with an aerial and then an interior view of the body/country, thereby passing the threshold between *terra cognita* and *incognita*, between the visible and invisible.[35] After a brief account of the country's etymology, situation, and topography, the focus shifts in Chapter 3, "Of *the* AIR, SOIL, RIVERS, CANALS, &c", from exterior to interior exploration. The remaining nine chapters of Pheuquewell's description of Merryland are exclusively dedicated to the country's interior, its animals, plants, and commodities, its provinces and "principal Places of Note" (*Merryland* 14) as well as some of its curiosities, before concluding his report with a mariner's guide on how to steer and anchor one's ship in Merryland harbor to enjoy "this delicious Country, without the Danger of *Waves, Tempests* or *Shipwreck*" (48).

The move from the exterior description of the female landscape's surface to the traveler's account of the country's interior is a decisive moment in the text. For one, it marks the traveler's crossing of the physical and visible boundaries of the country and his descent into the unknown. At the same time, the crossing is a moment of close bodily contact, reducing the physical distance between the observer and the observed. Pheuquewell describes the corporeal encounter between the male and female bodies with battle and siege rhetoric, emphasizing colonial metaphors while simultaneously negotiating sexual differences. Pheuquewell explains that prior to surveying the country's interior, the traveler has to pass two sets of "Forts" and "Fortresses" (15, 16).[36] However, the travel writer quickly adds that the female landscape's 'defense mechanisms' are "not very strong, tho' they have *Curtains, Hornworks,* and *Ramparts* [...] or never known to hold out long against a close and vigorous Attack" (15), and whereas some show "a stout Resistance, against strong Attacks and skilful Engineers", others "admit the Assailants without any Opposition" (16). The usage of battle and siege imagery and conquest rhetoric epitomizes some ideals of the heroic traveler and thus of heroic masculinity, male dominance, and power. The explorer is portrayed as brave, enduring, and vigorous, ready to attack and fight to complete his mission of (sexual) conquest. The passage demonstrates male domination and supremacy over women and their bodies and implies that women should either submit voluntarily to men's sexual urges or face sexual assault without consent.[37]

The concept of 'imperial rape' is later relativized in the text when Pheuquewell explains that Merryland, in fact, is a gynarchy "entirely under *Female Government*, there being an absolute Queen over each particular Province, whose Power is unlimited" (*Merryland* 32). This complication of the power dynamics in the text has been controversial. For Lewes, *Merryland* is an exclusive "metaphor of genital imperialism",[38] reinforcing the paradigm of imperial fantasy; Marcia Nichols, by contrast, criticizes Lewes's reading of the text and argues that "[w]omen have all the power in Merryland – and over their merrylands", and "men have to work to please their particular Queen, be she wife, mistress, or prostitute, or they risk being cast aside and replaced".[39] Lewes's discussion of somatopic erotica is insightful. Yet I agree with Nichols that denouncing "all somatic metaphors as imperial rape"[40] does not do justice to the multivalent gendered discourses addressed in the text.

Once inside the female body, the traveler's ocularcentrism is exchanged by haptic perception as the traveling I has to rely on his 'instrument' to continue his close observation and exploration of the female body. Comparable to a geographer, who in his attempt "to record information which cannot be actually *seen* [...] had to rely on indigenous informants who were treated as if they were themselves mechanical instruments",[41] the traveler's penis is turned into his most reliable informant and his 'new' organ of perception. What remains hidden from the traveler's eye, the traveling I can still explore and measure with his *"proper instrument"* (*Merryland* 4), described, for example, as a tool for navigation in ("Sounding-line") and cultivation of ("Plough") the interior of Merryland (11, 21). However, while objective methods of measuring are alluded to here, in contrast to "artificial technologies of vision"[42] used in the eighteenth century to underscore the mechanics of vision, the penis is a 'natural instrument of feeling' and consequently implies a 'feeling body'. Thus, despite the text's supposed focus on detached objectivity via observation, it nevertheless suggests touch. In the travelogue, phallocentric imagery is omnipresent. The text's obsession with the male sexual organ is pushed *ad absurdum* when it is described as a coral-like plant[43] and submarine animal, with no eyes yet sensible to touch, inhabiting the country's flora and fauna and thus part of the country's native environment, a natural inhabitant of the unknown territory (*Merryland* 25–26, 29–32). As such, *Merryland* is "a parody of the sexualized (and frequently sexist) language of the masculine genres of science, travel, and natural history".[44] And yet, although the text mimics the eroticized descriptions used by eighteenth-century geographers, botanists, and scientists, the text simultaneously reproduces them and thus continues in the same tradition.

Part of the ocularcentrism of the eighteenth century, technological devices to enhance vision, such as microscopes or binoculars, "established an almost physical distance between the viewer and the viewed, between the subject and the object of vision"[45] and thus detached the body from the experience. In *Merryland*, the traveling subject and his 'informant' form a symbiotic entity. Whereas sight allows the traveler to maintain a physical distance from the observed object, thus distancing the body and detaching the self from the experience, the penis not only reduces this distance but reverses it. The traveling I behind the 'eyes' is not disembodied from the experience but physically interacts with the woman-as-land – a contrast to the imperial eyes of the European colonizers, who, as Pratt argued, "seem powerless to act upon or interact with this landscape that offers itself".[46] As such, even if sight was the sense most commonly favored, here the travel writer's description of the country's interior continuously implies touch and human interaction between the male explorer and the female body-place, acknowledging the traveler's feeling body, its sensuality, and its possibility to experience (sexual) pleasure. Yet explicit references to the act of touching are rare and almost exclusively reserved for descriptions of both male and female masturbation. The male inhabitants of the country are said to be *"ticklish"* and like to *"tickle themselves"* (*Merryland* 19), while the queens are described to make use of "Manual Performances" resulting in "various Emotions and Agitations of the Body" (37). Whereas sight is linked to the male gaze and touch to sexual interaction and stimulation, the taste of the female body, by contrast, is completely desexualized and restricted to the "younger sort of People" who benefit from the nourishing and "wholesome Liquor" (29) produced by the breast.

The traveler's scientific expedition to Merryland takes the reader deep into women's reproductive bodies, "traditionally semanticized as dark, unchartered territory in the (male-)dominated cultural imagination".[47] These bodies, as the somatopic woman-as-land metaphor intriguingly shows, are portrayed as anonymous, as "interchangeable objects to conquer and cultivate"[48] and, generally, without history. In contrast to Pheuquewell's lengthy autobiographical account in the preface, the female body-place is presented as an everywoman, lacking any signs of individuality. The erasure of individuality allows the text to offer generalized knowledge about female bodies. At the same time, this juxtaposition of the traveling body's individuality and the female body-place's deindividualized uniformity mirrors the long-standing relation between men as subjects and women as objects. Moreover, it emphasizes that in "male narratives" women are often "the objects of desire or destination points rather than active co-travellers".[49]

In *Merryland*, the woman (and her body) is both an object of desire and a destination. Once the desired destination, the supposedly foreign female body-place, is reached, this 'exotic' object of desire, its territories, and commodities are 'cultivated', domesticized, and subsequently controlled by the successful colonist. In the tradition of explorative and colonial travel narratives, Pheuquewell repeatedly draws on agricultural metaphors to emphasize the fertility and suitability for husbandry of the female 'landscape', although such metaphorical references to agriculture "did not quite capture Enlightenment views of fruitful intercourse"[50] anymore. The narrator claims that, in particular, conquests promising the "first Tilling of a fresh Spot" are "most esteemed" (*Merryland* 11) and explains that all 'merrylands' are "generally fertile enough, where duly manured; and some Parts are so exceeding fruitful as to bear two or three Crops at a time" (12). These descriptions recall masculine fantasies about erotic womanhood commonly evolving around "virgin or maternal beauties"[51] while simultaneously echoing colonial narratives in which "new territories were metaphorized as female, as virgin lands waiting to be penetrated, ploughed, and husbanded by male explorers".[52]

2.3 Of "Unwholesome Provinces" and "Proper Clothing": Health, Travel Precautions, and Eighteenth-Century Medical Discourse

In his study of British colonial efforts in India, Edney remarks that "[t]he British engineer-surveyor looked at Indian landscapes as a surgeon looks at his patient, as an item to be thoroughly investigated, measured, and prodded so that maladies and imperfections might be identified, understood, adjusted, controlled, and so cured".[53] On Pheuquewell's travels to Merryland, the male, colonial, and medical gazes similarly merge when the perspective of the travel writer "transform[s] from a cartographic viewpoint into a position akin to a medical examination".[54] The image of the doctor–patient relationship in Edney's example recalls eighteenth-century medical endeavors. Central to the discourse was the premise that women and their bodies "could only be fully managed and understood through detailed objective, and professional learning".[55] Consequently, as Thomas Laqueur notes, "[w]omen's bodies in their corporeal, scientifically accessible correctness, in the very nature of their bones, nerves, and, most important, reproductive organs, came to bear an enormous new weight of meaning",[56] and their bodies were scrutinized and subjugated to an increasingly medicalized gaze.

Eighteenth-century scientific endeavors to describe newly discovered plants, animals, and places "involved all manner of linguistic apparatuses".[57] Parallel to the systematizing of the natural world and new territories, the scientific revolution caused a terminological transformation of genital nomenclature from metaphors connecting the reproductive body to nature to Latin, the language of science and elite male circles.[58] *Merryland* mirrors this decisive linguistic transition. Although the familiar allusions to the natural world are still used to explain a man and woman's primary and secondary sexual organs – the testes are portrayed as "*precious stones*" (24), pubic hair as "a spacious Forest" (17), for example – the description of the female landscape "is replete with the obvious modern genital landmarks".[59] Accompanied by "An Explanation of the Technical Abbreviations" (n. pag.), *Merryland* implements what was then relatively new scientific nomenclature for the female reproductive body. The reader is acquainted, for example, with "the little Mountain called MNSVNRS" (*mons veneris*), "two Forts called LBA" (*labia*), "the Metropolis, called CLTRS" (*clitoris*), and "a large Reservoir or Lake [...] called VSCA" (*vesica urinaria*), to give but a few examples (3, 15, 11). As Susan Bassnett points out, the process of systematizing and labeling was generally an "inherently male act", an attempt to order and control the world, which was the prerogative of the male intellectual elite.[60] In short, it was "[t]he (lettered, male, European) eye that [...] could familiarize ('naturalize') new sites/sights immediately upon contact, by incorporating them into the language of the system".[61] As such, using Latin terms reinforces the physical and intellectual superiority of the male traveler over women and their bodies.[62]

"To govern territories, one must know them";[63] therefore, to control and subsequently dominate newly discovered uncharted territory, colonizers needed to be aware of a country's geographical dimensions, document what they had seen, and transform their fields of work into maps that could then be used in the processes of colonization. Looking at and documenting a landscape, like Pheuquewell's exploration of Merryland, was not based solely on observation. Instead, it was a complicated construction of recollections and geographical imaginations, which intentionally or not reflected cultural, colonial, moral, and – in the case of *Merryland* – gendered interpretations of social relations and thus was "thoroughly ideological"[64] in its design. Pheuquewell refrains from providing a map "to give a more compleat Geographical Description of this Country, [...] recollecting it would considerably enhance the Price of the Book" (*Merryland* 18). Instead, he refers his readers to a medical chart of the female body, included in François Mauriceau's (1637–1709) anatomical atlas, *Les Maladies des Femmes grosses et accouchées* (1668), remarking that Mauriceau "was a great Traveller in that Country, and surveyed it with tolerable Exactness" so that by studying his 'map', "the Reader may see all the noted Places and Divisions laid down exactly as they are situated" (18).[65] Pheuquewell's recourse into anatomy exemplifies how imperialism, mapmaking, and anatomy intersect in their quest for knowledge and authority over territories (and women's bodies). Comparable to geographical maps, eighteenth-century anatomical atlases often described the human body in topographical terms.[66] In his preface, Mauriceau even assures his readers that his work is not, in the words of Harvey, "the medical equivalent of a travel lie".[67] The image Pheuquewell refers to features a female torso with two well-rounded breasts, intact pudenda, and carefully sketched pubic hair.[68] With the woman's viscera opened and the skin tissue peeled back, the "barriers between exterior and interior body"[69] are removed. This is a significant contrast to Pheuquewell's border crossing on his way to the interior of Merryland: the anatomist's map allows viewers to be "in direct scopophilic contact with the female form".[70] It permits us to look intently past the flesh into the body.[71] Furthermore, by selecting the posture and perspective of the object, highlighting and framing certain areas, and ignoring others, anatomists not only directed the viewer's gaze but, like the territorial explorers, also determined "those objects worth noticing and those to be passed over".[72] The travelogue's reference to anatomical atlases and its inclusion of scientific vocabulary alludes to the increasing medicalization, desexualization, and objectification of women's bodies prevalent in the eighteenth century and foreshadows the growing emphasis placed on eliminating any ambivalences about the female reproductive body to achieve control over it.

Next to its digression into medical and anatomical discourses and their visual representation and (linguistic) systematization, *Merryland* engages with eighteenth-century discourses on disease etiology, the threat of contagion, and the increasing impact climate was believed to have on the traveler's health. Pheuquewell notes that although the climate is generally quite pleasant, it sometimes is "so very hot, that Strangers inconsiderately coming into it, have suffered exceedingly; many have lost their Lives by it, some break out into Sores and Ulcers difficult to be cured; and others, if they escape with their Lives have lost a Member" (*Merryland* 8). Moreover, he warns future travelers repeatedly of the "extreamly [sic] gross and pestiential" (7) air in some provinces. The aspect of contaminated air is repeated in Pheuquewell's account of Merryland's flora and fauna. Here, the "Smell of the Air" is described as foul and fishy, and the sensual experience provokes a bodily response in the traveler – disgust – when he reports the putrid smell to be "very offensive" (23), which for a moment disrupts the text's erotic discourse.

As these examples show, the encounter with the other's body is thus not always positive and pleasurable for the traveler's own bodily experience; in fact, the 'fiendish climate' with its 'miasmic vapors' threatens the scientist-explorer's well-being and thus his mission. The descriptions of Merryland's climate and harmful air recall the miasma theory of disease etiology, which held that the transmission of contagious diseases, such as cholera and syphilis, was caused by different substances (such as water) or 'bad air' that infiltrated healthy bodies and made them sick. The country as a hazard to the traveler's health is brought up again when Pheuquewell refers to "*Blue* or *Roman Vitriol* (which is of great Use to eat away proud Flesh)", a mineral commodity to be "found on the Borders of this Country" (24). Vitriol (copper sulfate) was prescribed well into the nineteenth century for a variety of infectious diseases, including the plague, smallpox, scurvy, tinea capitis, and measles. It was either dissolved in liquid, such as water or tea, or applied as a tincture to all kinds of running sores.[73] Here, it underlines Pheuquewell's warning that "the Provinces, where this is found are generally unwholesome" (*Merryland* 24) and that, notwithstanding all bodily pleasure the exploration of Merryland may promise, the traveler should not be driven by sexual urges but rather should focus on reason to avoid the risk of turning from an active agent into a suffering patient. The examples semantically link the discourses of health and disease and reflect the interdependencies of both with travel, thus presenting another parallel to the genre of travel literature.

Moreover, these examples intriguingly show that diseases are presented as inherently female – it is women's bodies that are either transmitters or carriers of diseases. It is the country/the female body that is "so bewitchingly tempting [...] that People will too frequently rush into it without Caution, or Consideration of their Danger" (8), and make them forget the most significant characteristics of scientific explorers and colonists: (self-)control, discipline, reason, and close observation. For Pheuquewell, the encounter with the other's body can cause pleasure and pain. He, by contrast, distances himself from causing infections, contamination, and contagion in the 'provinces' he travels to and enjoys; he is either infected by the (female) landscape or there to cure it of any maladies and imperfections. His (sexual) encounter with the other's body is described as "very refreshing and nourishing" (32) and generally beneficial. It promises to restore "a most inexpressible resplendent Brightness to the whole Countenance" of the female body-place and rejuvenate it with "Life, Spirit, and juvenile Bloom" (25). *Merryland* reproduces the enduring belief that heterosexual intercourse could cure allegedly 'female conditions' like hysteria and chlorosis (green sickness), and early modern physicians frequently recommended and prescribed pelvic massages and sexual intercourse as a treatment against the supposed 'suffocations of the womb'.[74] In a similar tradition, Pheuquewell claims that sexual intercourse and notably the male semen "is reckoned a Specifick [sic] for the Green-sickness, and many other feminine Disorders; and is a

Medicine so wonderfully pleasant and easy in its Operation, that the nicest Palate or weakest Constitution may take it with Delight" (*Merryland* 32).

To cope with the "dangerous Heat of the Climate" (8) and to minimize the health hazards ascribed to "unhealthy Provinces" (9) and "dangerous Places" (44), *Merryland* informs readers on the contraceptive and protective methods available for both sexes, providing "sex education under the veil of allegory".[75] The traveler is advised "to wear *proper Cloathing* [sic]", described to be "very commodious" and "peculiarly adapted to this Country" (*Merryland* 9). The condom's probability of preventing pregnancies is not in focus here; instead, it is portrayed as a protective garment to safeguard the traveler's body from any unwanted contamination with sexually transmitted diseases, such as syphilis and gonorrhea, both endemic in Europe since the fifteenth century. Next to the condom, Pheuquewell also describes another common barrier contraceptive of the time to his readers: sponges soaked in vinegar used by women not only "as a *Cleanser*, but also as an Antidote against the bad Effects [pregnancy]" (26) conceivably caused by sexual intercourse.

Finally, despite its phallocentrism, patriarchal agenda, and focus on male pleasure and delight, it would be limiting to ignore the fact that *Merryland* indeed "celebrates female pleasure and orgasm"[76] – a stark contrast to the medical discourses of the time, which classified women as essentially passionless beings. The queens of Merryland, however, embrace their female sexuality. They, and they alone, take delight in their metropolis, pronounced as their "*Pleasure Seat*" (*Merryland* 15), and engage in sexual self-stimulation by using sex toys (36–37), which Pheuquewell ironically places in a chapter on religious practices, referring to it as "IMAGE WORSHIP" (36). By including references to female (self-)pleasure, erotic texts like *Merryland* challenged and possibly even "subverted the dominant ideology by creating space for positive female sexual expression outside the procreative norm".[77] Whereas eighteenth-century physicians dismissed the significance of vaginal lubrication for women's health and fertility and as an indicator of sexual arousal, *Merryland*, by contrast, makes use of various aquatic references. In addition to its river, canal, and floods, the air is described as "thick and moist" (7), and the country's soil is "naturally very wet and fenny", which "contributes much to its Fruitfulness" (10), as Pheuquewell explains. As such, moisture and water in their various materializations are significant tropes in the text, thus countering, or at least reacting to, this fundamental theoretical change of mentalities toward the female reproductive body.

2.4   Conclusion(s)

A New Description of Merryland demonstrates how travel writing conventions, erotica, and the Enlightenment's ideology behind knowledge production catalyzed to reproduce the century's concerns about sexual differences and "ways of thinking about and representing space – particularly cartography – that embodied gendered inequalities of power".[78] As such, it is more than an erotic male travel fantasy, but a text in which literary trends, science, and observation merge with erotic, colonial, and medical discourses, language, and imagery, inviting complex readings and presenting a highly gendered portrayal of the corporeality of traveling and of bodies in motion. With its fluid oscillation of geographical, botanical, agricultural, colonial, and scientific imagery, ideas, and nomenclatures, *Merryland*, as this chapter has shown, provides compelling insights into the multivalent and changing discourses on anatomy, sexuality, reproduction, and gender ideology emerging in the eighteenth century.

Notes

1 Roger Pheuquewell [pseud., alias Thomas Stretzer], *A New Description of Merryland: Containing a Topographical, Geographical, and Natural History of That Country*, 6th ed. (Bath: Printed and sold by J. Leake there; and by E. Curll, at Pope's Head in Rose-Street, Covent-Garden, 1741), pp. xiv–xv, emphasis in text. Further references to this edition are included parenthetically in the text. Unless otherwise noted, all emphases, italics, and capitalizations are in the original. Because Stretzer is only the alleged author, references will be made using the work's title (*Merryland*).
2 James Buzard, "The Grand Tour and After (1660–1840)", in Peter Hulme and Tim Youngs, eds., *The Cambridge Companion to Travel Writing* (Cambridge: Cambridge University Press, 2013), pp. 37–69 (at p. 38).
3 Stephanie Matos-Ayala, "British Essentialism in Eighteenth-Century British Travel Literature of the West Indies and North America", Ph.D. Thesis, University of Purdue, 2018, p. 1.
4 Mary Louise Pratt, *Imperial Eyes: Travel Writing and Transculturation* (London: Routledge, 1992), p. 29.
5 See Wortley Montagu's *Letters of the Right Honourable Lady M—y W——y M——e*, vols. 1–3 [*Turkish Embassy Letters*] (London: Printed for T. Becket and P. A. De Hondt, in the Strand, 1763), Smollett's *Travels through France and Italy*, vols. 1–2 (London: Printed for R. Baldwin, in Pater-Noster-Row, 1766), and Wollstonecraft's *Letters Written during a Short Residence in Sweden, Norway, and Denmark* (London: Printed for J. Johnson, St. Paul's Church-Yard, 1796). See also Buzard, "The Grand Tour", p. 37.
6 See Behn's *Oroonoko: or, the Royal Slave* (London: Printed for Will. Canning, at his Shop in the Temple-Cloysters, 1688), Defoe's *The Life and Strange Suprizing Adventures of Robinson Crusoe* (London: Printed for W. Taylor at the Ship in Pater-Noster-Row, 1719), and Swift's *Travels into Several Remote Nations of the World* [*Gulliver's Travels*], vols. 1–4 (London: Printed for Benj. Motte, at the Middle Temple-Gate in Fleet-Street, 1726). See also Karen Harvey, *Reading Sex in the Eighteenth Century: Bodies and Gender in English Erotic Culture* (Cambridge: Cambridge University Press, 2005), p. 176.
7 Percy G. Adams, *Travelers and Travel Liars 1660–1800* (Berkeley and Los Angeles: University of California Press, 1962), p. 9. See also Buzard, "The Grand Tour", p. 37.

8 Harvey, *Reading Sex*, p. 175.
9 [Charles Cotton], *Erōtopolis, the Present State of Betty-Land* (London: For Tho. Foy, at the White-Hart, over and against St. Dunstan's Church in Fleet-Street, and at the Angel in Westminster-Hall, 1684). For more on *Erōtopolis*, see Julie Peakman, *Mighty Lewd Books. The Development of Pornography in Eighteenth-Century England* (Basingstoke and New York: Palgrave Macmillan, 2003), pp. 97–100; and Harvey, *Reading Sex*, pp. 103–5.
10 Harvey, *Reading Sex*, p. 2.
11 *Records of the Most Ancient and Puissant Order of the Beggar's Benison and Merryland, Anstruther*, reprinted, ed. Alan Bold (Anstruther, 1892), pp. 6–7, emphasis in text. *Merryland* includes an allusion to the Order when stating that "over a Bottle [...], they begin with drinking a Health to MERRYLAND" (*Merryland* 21), which points to its tradition to recite licentious toasts at its gatherings. See also Franz Reitinger, "Mapping Relationships: Allegory, Gender and the Cartographical Image in Eighteenth-Century France and England", *Imago Mundi*, 51 (1999): 106–30 (at p. 118); and Harvey, *Reading Sex*, pp. 62–63.
12 Reitinger, "Mapping Relationships", p. 118.
13 See also Reitinger, "Mapping Relationships", p. 118.
14 Other 'Merryland' texts include, for example, Philo-Brittanniae [pseud.], *The Potent Ally: or, Succours from Merryland. With Three Essays in Praise of the Cloathing of That Country and the Story of Pandora's Box. To Which Is Added Erōtopolis. The Present State of Bettyland* ('Paris' [London]: Printed by direction of the author, and sold by the booksellers of London and Westminster, 1741); [Thomas Stretzer], *Merryland Displayed: or, Plagiarism, Ignorance, and Impudence, Detected* (Bath [London]: Printed for the author, and sold by J. Leake; and the booksellers of London and Westminster, 1741); and *A Compleat Sett of Charts of the Coasts of Merryland* (1745). No copies are extant of the latter (see Harvey, *Reading Sex*, p. 37). For similar texts, see also Captain Samuel Cock [pseud.], *A Voyage to Lethe; By Capt. Samuel Cock; Sometime Commander of the Good Ship the Charming Sally. Dedicated to the Right Worshipful Adam Cock, Esq; of Black-Mary's-Hole, Coney-Skin Merchant* (London: For J. Conybeare in Smock-Ally near Petticoat Lane in Spittlefields, 1741), in which the female body is portrayed as a ship.
15 From Greek 'soma' (body) and 'topos' (space).
16 Darby Lewes, "Utopian Sexual Landscapes: An Annotated Checklist of British Somatopias", *Utopian Studies*, 7, no. 2 (1996): 167–95 (at p. 167, emphasis in text).
17 Karen Harvey, "Gender, Space and Modernity in Eighteenth-Century England: A Place Called Sex", *History Workshop Journal*, 51 (2001): 159–80 (at p. 174).
18 This does not mean women did not read and possibly enjoyed the work. In *Merryland Displayed* (1741), Stretzer complains that he is "sorry to find that some of the Fair-Sex, as well as the Men, have too freely testified their Approbation of this *pretty* Pamphlet [...] and that over a Tea-Table some of them make no more Scruple of mentioning *Merryland*" ([Stretzer], *Merryland Displayed*, p. 5, emphasis in text). For more on women and the erotic book trade, see Peakman, *Mighty Lewd Books*, pp. 25–26.
19 Marcia Nichols, "Roger Phequewell, Colonial Man of Science: Re-Reading Imperial Fantasy in Merryland", in Judy A. Hayden, ed., *Travel Narratives, the New Science, and Literary Discourse, 1569–1750* (London and New York: Routledge, 2012), pp. 143–57 (at p. 149).
20 Harvey, *Reading Sex*, p. 176.

21 Thomas Laqueur, *Making Sex: Body and Gender from the Greeks to Freud* (Cambridge: Harvard University Press, 1992), p. 159.

22 For an analysis of the subordination of women to an increasingly male-dominated medical discourse, see, for example, Susan C. Staub, "Surveilling the Secrets of the Female Body: The Contest for Reproductive Authority in the Popular Press of the Seventeenth Century", in Andrew Mangham and Greta Depledge, eds., *The Female Body in Medicine and Literature* (Liverpool: Liverpool University Press, 2012), pp. 51–68; Marcia D. Nichols, "Venus Dissected: The Visual Blazon of Mid-Eighteenth-Century Medical Atlases", in Jolene Zigarovich, ed., *Sex and Death in Eighteenth-Century Literature* (New York: Routledge, 2013), pp. 103–23; Wieland Schwanebeck, "The Womb as Battlefield: Debating Medical Authority in the Renaissance Midwife Manual", *Journal for the Study of British Cultures*, 23, no. 2 (2016): 101–14; and Sarah Schäfer-Althaus, "Dissecting Birth: Obstetrics, Bodily Transience and the Anatomist's Gaze in Eighteenth-Century Medical Atlases", in Sarah Schäfer-Althaus and Sara Strauß, eds., *Transient Bodies in Anglophone Literature and Culture* (Heidelberg: Universitätsverlag Winter, 2020), pp. 23–46.

23 Schäfer-Althaus, "Dissecting Birth", p. 23. See also Laqueur, *Making Sex*, p. 154.

24 William H. Sherman, "Stirring and Searchings (1500–1720)", in Hulme and Youngs, eds., *Companion to Travel Writing*, pp. 17–36 (at p. 30).

25 Harvey, *Reading Sex*, p. 176.

26 Harvey, *Reading Sex*, p. 176.

27 For nationalism as a common motif in erotic culture see Harvey, *Reading Sex*, pp. 139–45.

28 Nichols, "Roger Phequewell", p. 149.

29 This also indicates the century's insufficient technological means to define geographical coordinates precisely (Matthew Edney, *Mapping an Empire: The Geographical Construction of British India, 1765–1843* (Chicago: University of Chicago Press, 1999), p. 39).

30 Edney, *Mapping an Empire*, p. 48.

31 "Insight: Search Online Etymology Dictionary", Etymology, https://www.etymonline.com/search?q=insight [01 December 2022].

32 Edney, *Mapping an Empire*, p. 48.

33 In fact, Pheuquewell, the observer-traveler, ridicules the spectator-travelers and their failure at visual comprehension and evidence-based reasoning by alluding to the Mary Toft-affair (1726): "I know it has been strongly insisted on by several learned Men (some of them great Travellers in MERRYLAND that *Rabbets* have been bred in that Country, and they expected great Profits [...] but, after a great Noise made about it, All came to nothing" (*Merryland* 24). Toft claimed to have given birth to rabbits after a particular appetite for their meat during pregnancy. A public sensation, she was examined by leading specialists in the field. The discovery of the hoax resulted in long-lasting ridicule of the medical men involved. For a contemporary account of the case, see Nathaniel St. André, *A Short Narrative of an Extraordinary Delivery of Rabbets, Perform'd by Mr. John Howard, Surgeon at Guilford* (London: John Clarke, 1727). See also S.A. Seligman, "Mary Toft – The Rabbit Breeder", *Medical History*, 5, no. 4 (1961): 349–60.

34 Pratt, *Imperial Eyes*, p. 23.

35 See also Harvey, *Reading Sex*, p. 177.

36 The "forts" are a reference to the *labia majora*, the "fortresses" to the *labia minora* (*nymphae*).

37 See also Darby Lewes, "Nudes from Nowhere: Pornography, Empire, and Utopia", *Utopian Studies*, 4, no. 2 (1993): 66–73 (at p. 66).

38 Lewes, "Nudes from Nowhere", p. 66.

39 Nichols, "Roger Phequewell", p. 156. See also *Merryland* 32–34 for the detailed description of its queens.

40 Nichols, "Roger Phequewell", p. 157.

41 Edney, *Mapping an Empire*, p. 51.

42 Edney, *Mapping an Empire*, p. 48.

43 At the time *Merryland* was published, corals were still classified as plants. It was not until the late eighteenth century that Sir Frederick William Herschel, Fellow of the Royal Society, discovered with the help of a microscope that corals were, in fact, animals.

44 Nichols, "Roger Phequewell", p. 157.

45 Edney, *Mapping an Empire*, p. 48.

46 Pratt, *Imperial Eyes*, p. 60.

47 Schwanebeck, "The Womb as Battlefield", p. 105.

48 Nichols, "Roger Phequewell", p. 143.

49 Susan Bassnett, "Travel Writing and Gender", in Hulme and Youngs, eds., *Companion to Travel Writing*, pp. 225–41 (at p. 225).

50 Laqueur, *Making Sex*, p. 154.

51 Nichols, "Roger Phequewell", p. 147.

52 Bassnett, "Travel Writing and Gender", p. 231.

53 Edney, *Mapping an Empire*, p. 52–53.

54 Harvey, *Reading Sex*, p. 177.

55 Sheena Sommers, "Transcending the Sexed Body: Reason, Sympathy, and 'Thinking Machines' in the Debates over Male Midwifery", in Andrew Mangham and Greta Depledge, eds., *The Female Body in Medicine and Literature* (Liverpool: Liverpool University Press, 2012), pp. 89–106 (at p. 89).

56 Laqueur, *Making Sex*, p. 150.

57 Pratt, *Imperial Eyes*, p. 29.

58 See Laqueur, *Making Sex*, p. 154.

59 Laqueur, *Making Sex*, p. 160.

60 Bassnett, "Travel Writing and Gender", p. 230.

61 Pratt, *Imperial Eyes*, p. 31.

62 Bassnett, "Travel Writing and Gender", p. 231; see also Harvey, *Reading Sex*, p. 57.

63 Edney, *Mapping an Empire*, p. 1.

64 Edney, *Mapping an Empire*, p. 52.

65 Next to Mauriceau's map, Pheuquewell, however, admits that the "*Model or Machine*" by Richard Manningham ("Sir R.M.") provides an even "better Idea of MERRYLAND than can possibly be done by the best Maps, or any written Description" (*Merryland* 18) – a reference to the three-dimensional obstetric simulators known as 'artificial matrix' used by obstetricians to teach students obstetric maneuvers and complicated deliveries while simultaneously preventing any physical harm of pregnant women by "Pupils practicing too early on real Objects", as Manningham stated in an advertisement published in the *London Evening Post* (1740) ("From the Lying-in Infirmary in Jermyn-Street, St James", *London Evening Post* (01 March 1740)).

66 Schäfer-Althaus, "Dissecting Birth", p. 30.

67 Harvey, *Reading Sex*, p. 184. As Mauriceau states: "I would not hinder your reading other learned Authors who treat of it, but only acquaint you, that the most part of them (having never practis'd what they undertake to teach)

resemble [...] those *Geographers,* who give us the Description of Countries they never saw, and (as they imagine) a perfect Account of them" (François [Francis] Mauriceau, *The Diseases of Women with Child, and in Child-Bed: As Also the Best Means of Helping Them in Natural and Unnatural Labour,* 7th ed., trans. Hugh Chamberlen (London: For T. Cox, at the Lamb, and J. Clarke, at the Bible, under the Royal-Exchange in Cornhill, 1736), pp. vii–viii, emphasis in text).

68 Mauriceau, *The Diseases of Women,* p. 24. The translator, Hugh Chamberlen, was himself an esteemed accoucheur coming from a family of surgeons who invented the obstetric forceps. Pubic hair was a significant element in erotic literature and, together with the breast, was regarded as "the most visible and traditionally eroticised symbols of femininity, female sexuality and sexual maturity" (Schäfer-Althaus, "Dissecting Birth", p. 33). For an analysis of pubic hair in eighteenth-century anatomy atlases, see Nichols, "Venus Dissected", pp. 105–12; and also Harvey, *Reading Sex,* p. 97.

69 Schäfer-Althaus, "Dissecting Birth", p. 33.

70 Laura Mulvey, "Visual Pleasure and Narrative Cinema", in Gerald Mast and Marshall Cohen, eds., *Film Theory and Criticism* (New York: Oxford University Press, 1975), pp. 803–15 (at p. 811).

71 Harvey, *Reading Sex,* p. 184.

72 Edney, *Mapping an Empire,* p. 49.

73 See, for example, John Wesley, *Primitive Physick; or an Easy and Natural Method of Curing Most Diseases,* 9th ed. (Holborn, London: Thomas Trye, 1761), p. 90; and William Buchan, *Domestic Medicine, or, a Treatise on the Prevention and Cure of Diseases, by Regimen and Simple Medicines,* 22nd ed. (Boston: Otis, Broaders, and Company, 1848 [1769]) pp. 173–74, 178, 283, 429.

74 The French surgeon Ambroise Paré (1510–1590), for example, proposes that "if shee bee married, let her forthwith use copulation, and bee strongly encountered by her husband, for there is no remedie more present then this" (*Works of That Famous Chirugion,* trans. Thomas Johnson (London: Richard Cotes, 1634), p. 634). Similar ideas are expressed throughout the eighteenth century. See, for example, Nathanial Highmore's *De Passione Hysterica et Affectione Hypochondriaca* (1660) and Bernard Mandeville's *Treatise of the Hypochondriack and Hysteric Passions* (1711), which recommended horse-back riding and lengthy pelvic massages to induce hysterical paroxysms. By the nineteenth century, 'hysterical' women were regarded as a lucrative source of income in England's medical market.

75 Reitinger, "Mapping Relationships", p. 118.

76 Nichols, "Roger Phequewell", p. 156.

77 Nichols, "Roger Phequewell", p. 157.

78 Harvey, *Reading Sex,* p. 177.

3 From Facts to Physicality

Body Concepts in German Travel Writing Around 1800

Sonja Klein

3.1 Introduction

How disgusting I find my descriptions when I reread them! Only your advice, your bidding, your command can induce me to do so. Then too, I had read so many descriptions of these objects before I ever saw them! And did they afford me a picture of them, or even the faintest notion? In vain did my imagination labor to produce these objects, in vain did my mind strive to get some idea from them. Now I stand and contemplate these marvels, and what is the effect upon me? I have no thoughts, I have no feelings [...].[1]

It is indeed a peculiar opening that Johann Wolfgang von Goethe chose for his still little-known *First Part* of the *Letters from Switzerland* (*Briefe aus der Schweiz. Erste Abtheilung*, 1808). In its supposed author, Werther, we not only encounter a post-mortem character who the poet had already killed off spectacularly in his first novel (*The Sorrows of Young Werther*; *Die Leiden des jungen Werthers*) in 1774 and whose prehistory he now, somewhat surprisingly, presents 30 years later. We also meet a travel writer who consistently circumvents all of the reader's expectations of such a text genre from the onset. The title, *Letters from Switzerland. Part the First*, seems to promise contemporary observations about the country being visited, fleshed out by descriptions of the places seen, their inhabitants, their customs, traditions, and the natural spectacles, known to readers from countless other travelogues and letters typical of the age. Werther, however, announces at the very beginning that he is neither willing nor able to deliver such an account. Surrounded by the wonders of Switzerland's natural landscape, he is rendered bereft of 'Einbildungskraft' [imagination]; instead of the immensity and magnificence of the scenes he is confronted with, he is filled only with an absolute emptiness: "I have no thoughts, I have no feelings" (Goethe, *Letters* 1). This is all the more astonishing when one recalls how the Werther in Goethe's debut novel was a figure who, when it came down to it, defined himself almost solely through his 'Empfindsamkeit' [sensibility] and feelings.

DOI: 10.4324/9781003331803-4

Consequently, the reader knows not to expect a 'sentimental journey' in the tradition of Laurence Sterne. Such an expectation is just as abruptly shattered by Werther's first letter and its equally abrupt beginning as – and this is indeed surprising – the choice of the genre of travel literature in itself is radically questioned. Posing the rhetorical question, "I had read so many descriptions of these objects before I ever saw them! And did they afford me a picture of them, or even the faintest notion?" (1), Werther reveals himself to be not only a diligent reader of the already available travel literature about Switzerland but also a person who remains cut off from any personal connection to the country he is traveling because of these very pretexts and reading material. In this, he seems to be a 'victim' of what Ansgar Nünning has termed 'prefiguration' or 'premediation' ["Präfiguration"].[2] Accordingly, what Werther has read about Switzerland and its sights does not contribute to any increased attentiveness or local knowledge during his tour but causes rather the opposite: it leads to disorientation, the lack of imagination, and, resulting from this, finally to a crisis of language.

The mere fact that Werther's letters are provided without dates or locations shows how the writer distances himself from all specification and facts, without which a travel narrative at the time was normally unthinkable.[3] In the 15 *Swiss Letters*, hardly a single place is named nor a location described. In addition, Werther refrains from providing names of the people he has encountered on his trip apart from giving anonymous initials. When he writes at the beginning, "Now I stand and contemplate these marvels" (Goethe, *Letters* 1), the reader knows neither where exactly Werther is writing from nor which Swiss "marvels" he is contemplating. On the rare occasions when a place is actually named, and this happens only twice in the text, the descriptions that the reader might reasonably expect are left out completely. Werther laconically mentions, as if in passing: "We came to Geneva" (13). And only little more is imparted in his description of his trek up the St. Gotthardt: "Yes, I have climbed the Furka Pass and the St. Gotthardt Pass. These sublime, incomparable scenes of nature will always be present to my mind" (9). The reader, however, is presented with none of this imagery, the grandeur of these "scenes of nature" remains withheld from him as its very intensity makes it 'inexpressible'[4] for Werther and thus indescribable from his standpoint. Consequently, his letters can be seen as a kind of deficient travel report which merely circles around the *impossibility* of description, of telling, of reporting, and of portraying what he sees and experiences: "what can I, what do I do? Why, I sit down and scribble and describe. Go on your way, then, descriptions! Delude my friend, make him believe that I am doing something, that he is seeing and reading something" (1). But neither the addressed "friend" nor the contemporary reader of Goethe's text would be deluded and instead recognize from the start that these peculiar *Letters from Switzerland* were anything but a travel report.

In my essay I will analyze these *Swiss Letters*, placing them in the context of other travelogues and texts by Goethe as well as of German eighteenth-century travel literature in general. They will be interpreted as a reaction to an important paradigm shift around 1800, which lead to a new perspective on the human body and also strongly influenced the genre of travel writing.

In my study, the body is understood as a kind of construction that is influenced not mainly by the body's physiological characteristics but its cultural codification.[5] Body concepts have been subject to continual change over the centuries and at all times. In this constant metamorphosis, they appear to be primarily the result of "the history of the body's perception and its emblematic intermediation"[6] and have to be situated "at the intersection of discourse and materiality".[7] In the following, I argue that Goethe had an important part in shaping this discourse around 1800 and that his writing was of impact not only for the generations to come but also for the emergence of a new kind of body-conscious traveler.

3.2   Facts and Figures

Notwithstanding Werther's shortcomings as a travel writer in the first part of *Letters from Switzerland*, Goethe himself was well aware of what contemporary readers of a travel diary or travel literature *would* be expecting. Not only did he deliver a travel report perfectly adjusted to the taste and fashion of the time in his popular *Letters to Switzerland. Part the Second*,[8] in which he combined a detailed account of the route through Switzerland, giving all the necessary data, with a description of the effects various destinations and sights have on their visitor. He also experimented with the genre already in his youth, as his father, Johann Caspar Goethe, repeatedly reminded him to keep a journal while touring. The latter made a considerable contribution to the fashionable genre himself, fastidiously and extensively detailing his 'Grand Tour' through Germany, Austria, Italy, and France in the year 1740 in a series of fictitious letters written in Italian. His *Viaggio per l'Italia* is a typical document of the age – focusing on the clarity of its depiction of what contemporaries considered necessary facts and figures and compiled as one of those much-touted postulates of usefulness which characterized the 'learned travel' report of the early and middle years of the Age of Enlightenment.[9] This encyclopedic interest, which *Zedlers Universallexikon* defined in its much-quoted article "Reisen" [Traveling] of 1742 as "[t]he intended purpose of traveling is to get to know the world, and that means that one observes nations in their customs, morals, and manners, and employs this useful knowledge for one's own benefit",[10] also characterizes the Italian letters by

Goethe's father. To ensure the most detailed and fact-filled composition of his *Viaggio per l'Italia*, Johann Caspar Goethe drew not only on his own experience and memories but also on already existing travel guides, which he quoted extensively.[11]

The extent of these Enlightenment-inspired efforts to produce informative 'apodemics' (instructive travel literature) is demonstrated in Friedrich Nicolai's *Beschreibung einer Reise durch Deutschland und die Schweiz im Jahre 1781*. Published between 1783 and 1796, it represents (in terms of its magnitude, if nothing else) the highpoint as well as the end of academic travel reporting as a genre within German literature. This practically unreadable mammoth work – packed with descriptions that lose themselves in minutiae – fills an alarming 12 volumes and yet remains an unfinished fragment. Despite its meticulousness, however, it seems curiously lacking any system or logic – as is inevitable in any work striving to include the totality of desired and 'useful' knowledge. Figures and statistics fill row upon row of text; registers of births and deaths follow geological, mercantile, and political tables and lists. In the sixth volume of his travel lexicon, Nicolai has still not even reached Switzerland and makes a "small" yet voluminously described "side trip from Vienna to the border of Hungary and back to Vienna again".[12] And indeed, the *Jenaer Allgemeine Literatur-Zeitung* even praised the work in 1786 with the words: "Only few travel reports are as instructive and composed with such a noble candour and such an impartial love for truth as Nicolai's."[13]

However, around 1800, these kinds of publications were already losing their appeal and gave way to a new kind of travel literature. The change in the perception of travel as in the production of travel literature is mirrored, amongst others, in Heinrich Heine's beloved *Travel Pictures* (*Reisebilder*; 1826). Here the author – at the onset of his text *The Harz Journey* (1824) – caricatures the mania for facts prevalent in the eighteenth century by a hilarious description of the alleged excessive size of the feet of the ladies in Göttingen, using a satirically exaggerated 'apodemics' style filled with pseudo-statistics:

> Fuller information about the town of Göttingen may be conveniently obtained from the topography of the same by K.F.H. Marx. Although I am under the most sacred obligations to the author [...], I cannot recommend his work unreservedly, and I must deplore his failure adequately to rebut the erroneous view that the women of Göttingen have excessively large feet. Indeed, I have been at work for many a long day on a serious refutation of this view, and have therefore attended lectures on comparative anatomy, taken notes from the rarest books in the library, and spent hours studying the feet of the ladies passing along Weender Street. In the work of profound learning which will embody the results of these studies, I discuss: 1) feet in general,

2) feet in the ancient world, 3) elephant's feet, 4) the feet of Göttingen women, 5) I collect everything that has been said about these feet in Ulrich's Beer Garden, 6) I examine these feet in their context, and add an extensive discussion of calves, knees, etc., and finally 7) if I can find any paper large enough, I shall add some copperplate engravings with facsimiles of the feet of Göttingen ladies.[14]

Besides the satirical brilliance, with which this passage mocks the learned travel reports of previous decades, in its emphasis on physicality (feet, calves, knees, etc.) it also indicates the significant change in travel writing around 1800, which focused less and less on facts and figures, while becoming increasingly engaged with the individuality of the traveler himself and thus with the physicality of her/his bodily movement in space.

While, for a long time, the main goal of traveling had been the statistically measurable acquisition of knowledge about foreign places, people, and culture, the new priority was to round out the traveler's personality and his emotional "Bildungsweg" [course of education]. In tune with the momentous developments within the perception of art and literature in the late eighteenth century, during which sensitivity and the genius aesthetics of the younger generation of German writers superseded the strictly regulated poetry of previous centuries, the descriptions of visited locations now paid far less attention to the rules, tables, and statistics and more to the emotions they aroused within the subject himself. The Enlightenment challenge of 'discovering the world' became a voyage of self-discovery, a development of the writer's internal *terra incognita*, and in this, a key goal particularly for German Romanticist writers. Thus, Novalis states in his sixteenth *Blüthenstaub-Fragment*: "We dream of traveling through the universe – but is not the universe *within ourselves*? The depths of our spirit are unknown to us – the mysterious way leads inwards."[15]

This prioritization of the individual, interior, and physical experience now meant that the importance of a location was dependent on its subjective effect on the traveler and thus on her/his body. Consequently, the focus of travel gradually shifted from an outside to an inside world, from objective facts to subjective physicality, and in this shift also mirrored the anatomical zeal of an age that had set out to re-discover the human body and its yet unknown depths in the arts as well as in science: the late eighteenth century saw the emergence of modern anthropology, with its many different disciplines growing into specialized areas. Via technical and medical innovations, anatomy and the dissection of bodies were not only considered the "highpoint of progress",[16] which led to countless new anatomical discoveries, but also regarded as an instructive entertainment for a lay audience.[17] Medical education was improved and modernized, and numerous publications and new periodicals engaged with all kinds of aspects relating to the human body and soul. The arts reacted in a similar way to this 'physical revolution'. Thus, the literary discourse in an age that Torsten Hoffmann even calls the "Jahrhundert des Körpers"[18] [Century of the Body] is intimately linked to the examination, exploration, and invention of creative strategies which re-inscribe the body in order to re-describe the world. On the whole and especially compared to previous centuries, the eighteenth century marks a decisive shift in terms of body concepts and subjectivity:

> The medieval subject was occupied with 'reading' the book of nature for which she/he represented a minute but nevertheless integrated part of text, that is, she/he was occupied with the hermeneutics of existence. As a representational medium, she/he saw the body as allegory of the world. In contrast, the subject of the 18th century discovered the body and its parameters as a heuristic cognitive tool providing the capability to construct, lead and influence imagination, a new narrative, or new connections of understanding regarding the inner and outer world, thought and matter, in other words, insight, carried by discourse networks, of which no one had been aware before.[19]

3.3 Physical Movements

One of the first German writers to literarily react to these developments and (re-)discover the body and its potentials was Goethe, who in his early lyrical concept of the 'Wanderer' (see for example *Wanderers Sturmlied*, 1772; *Wanderers Nachtlied*, 1776; *Harzreise im Winter*, 1777), created a new kind of body-conscious traveler, which would also shape his later prose texts. Via meter, enjambment, assonance, and other phonetic devices, these poems reflect the rhythm of the quickened breath and pace of the wandering speaker and recreate his physical movements in space with words and sound.[20] Thus, the poems become an essentially corporeal experience not only for the lyrical speaker but also for the reader who recites the poem and in reading out loud adapts his own breath to that of the traveler. Furthermore, in rejecting other devices of transport and deliberately choosing pedestrian travel, the 'Wanderer' seeks to physically amalgamate with the landscapes and places he traverses, until the traveling body and nature seem to merge into one.[21]

The fact that the transition from the eighteenth to the nineteenth century marked a radical as well as a physical change for travel literature is likewise marked in Goethe's epoch-making Bildungsroman *Wilhelm Meister's Apprenticeship* (*Wilhelm Meisters Lehrjahre*, 1796). While traveling with a theater troupe, the book's protagonist finds out that actual knowledge of the world is no longer acquired via facts and figures. Asked by his father to deliver a report on his travels in a "tabulary scheme" that will provide an account of his journey, Wilhelm at first promises his family "a copious journal of his travels, with all the required geographical, statistical, and mercantile remarks".[22] As, however, he is about to start he realizes that although he could write about his "emotions and thoughts, and many experiences of the heart and spirit", he could not do the same for concrete facts relating to "outward objects, on which, as he now discovered, he had not bestowed the least attention".[23] Thus, in the end, Wilhelm Meister's years of apprenticeship will not be completed via collected 'useful information' and data about the visited locations but through his detailed observation of the *eloquentia corporis* of the actors he is traveling with, and – most of all – his own bodily experiences of love, sickness, physical injury,[24] and acting. Consequently, in the later *Journeyman Years* (*Wilhelm Meisters Wanderjahre oder Die Entsagenden*, 1821–29), Wilhelm will retrospectively sum up his theatrical travels in the statement that "all things considered, physical man plays the main role in that world".[25]

This focus on the physical also holds the key to understanding the aforementioned *Letters from Switzerland. Part the First.* Here, however, the (traveling) body figures in such a prominent way that the text did not only provoke a veritable scandal at the time of its first publication and shocked Goethe's contemporaries to such a degree that hardly anybody would even mention it. It also remains to this day 'one of Goethe's least known poetical works' ["eines der unbekanntesten poetischen Werke Goethes"].[26]

But what happened in these strange travel letters that affronted readers and Goethe scholars of various generations? After failing in all his attempts to put his impressions of Switzerland into words, Werther turns his back on the geographical aspects of the area and focuses instead on the human body. Here at last, it seems, he stumbles across new territory. For example, when an unnamed Swiss "art-lover" goes to a secret cabinet and fetches out an erotic painting of Danae receiving the golden rain of Zeus in her lap, Werther's aesthetically trained eye allows him to effusively spout academic platitudes about "the splendor of the limbs, [...] the magnificence of situation and posture, [...] the grandeur of the tenderness and [...] the brilliance of that most sensual object" (Goethe, *Letters* 13). He himself, however, remains – as before when facing the wonders of the Swiss landscape – cold and unmoved: "I only stood before it in contemplation. It did not arouse *that* rapture, *that* joy, *that* inexpressible delight in me" (13). On the contrary, the sight of the painting makes him "uneasy", as it becomes clear to him that

> of that masterpiece of Nature, the human body, of the connection and harmony of its structure, I have [...] no conception at all. My imagination does not present this glorious edifice to me in a vivid way, and

when art presents it to me, I am in no position either to feel anything
as I behold it.

(13)

Comparable to the complaint in the opening letter, Werther is again left
incapable of both thinking and feeling. To overcome "this dull condition"
and to get to know "that masterpiece of nature, the human body" (13), he
shortly afterwards invites his traveling companion Ferdinand to go swim-
ming naked in a Swiss lake; an activity which he observes extensively and
with voyeuristic fervor. Admittedly, this bathing experience does elicit an
interest in Werther for the beauty of the naked male form, but, again, it
results in a purely theoretical aesthetic evaluation that does not include
individual feeling or sensuality: "how admirably formed is my young
friend! what a symmetry in all his limbs! what fullness of form, what a
glory of youth! and what a gain for me, to have enriched my imagination
with this perfect model of human nature!" (13). Werther's eulogy to his
friend's body does not concentrate on the vitality of the individual, but
rather the "perfect model" ["vollkommenes Muster"] which is praised in
much the same way as one would admire a fine antique statue. Ferdinand's
anatomy is perceived as an idealized, as it were, inflexible archetype of
the flawless human body which corresponds to an abstract and elusive
concept of 'beauty' as propagated in the influential artistic contemplations
of the art historian Johann Joachim Winckelmann (*Gedancken über die
Nachahmung der Griechischen Wercke in der Malerey und Bildhauer-
Kunst*, 1755). In this passage, Werther also shows himself to be a person
whose earlier reading (here the writings of Winckelmann and his succes-
sors) precludes him from any individual or spontaneous interpretation of
what he has viewed. Like the travel guides read before Werther's visit to
Switzerland, the 'art guide', another medium of writing, again leads to preconceived notions that make any authentic experience impossible for him.

It is thus hardly surprising that Werther's last travel destination, the
female body, also results in disappointment. The text ends – and this was
what made it so scandalous in the eyes of his contemporaries – with a visit
to a prostitute in Geneva, who Werther selects as the model for his studies of the female form.[27] While the young woman is artfully undressing
in front of Werther he, once more, proves to be an eager observer, who
nevertheless cannot be brought to feel anything. Although he does not only
consider the woman "beautiful" (Goethe, *Letters* 14) but is also openly
offered sexual intercourse by her, Werther maintains a safe distance and
restricts himself to voyeuristic viewing only. Again, he fails to either overcome his reserved attitude or involve his feelings. Even when confronted
with this last Swiss "wonder of nature" the traveling tourist is left cold.
In the end, shortly before the text breaks off, the naked female body even

seems "strange" to him and triggers "an awe-inspiring impression": "Shall I confess it: I could as little reconcile myself to that glorious body, when the last covering fell, as perhaps our friend L.*[28] will be able to be reconciled to his state, if Heaven should make him the Chieftain of the Mohawks. What do we see in women?" (15).

By now, after the body of a woman is compared not only with a Native American tribe, but also linked in the same breath with the foreign and "strange" that seems to be threatening, even the most obliging of readers must realize that these *Swiss Letters* can be nothing other than a parody. In his images of nakedness, so provoking for the contemporary reader, his often sexually connotated wordings, his concrete references to the body and, not least, his merciless deconstruction of one of the most beloved characters of German literature (Werther), Goethe harshly criticizes the standard genre of the learned travel report as well as a purely rational and scientific approach to the world. By representing Werther as a man of missed opportunities, by not allowing him to grasp or feel anything, and by repeatedly confronting him with the bondage of his own intellect and the preconceived notions which he has adopted not from individual experience but from reading books (Swiss travel reports, Winckelmann's studies, etc.), Goethe delivers a clear rejection of Enlightenment 'apodemics' and at the same time a refusal of art that would not combine sense *and* sensibility.

For Werther, the result of his exclusive concentration on voyeuristic viewing and his disconnection of the sensuality of his own body led him to artistic impotence and a distanced attitude towards the country he is traveling through, the Danae image, the naked bather Ferdinand, and lastly the young prostitute. By only relying on rational elements, he excludes all aspects of the physical so necessary for an individual experience. Cut off from his own body as well as those of others, he remains a cold, intellectual observer, who in the end sees nothing at all.

3.4   Sense and Sensibility

Goethe's aesthetic counterprogram to this parody of corporeal failure is found in his *Roman Elegies* (*Römische Elegien*, 1788/90), at first titled *Erotica Romana*, where the reader encounters a poetic and, in physical terms, veritably potent lyrical traveler through Italy. Coming to Rome for the first time, the speaker emphasizes from the very beginning of the *Elegies* that getting to know Rome does not only mean to read about it and visit its sights, but also to sensually feel it through the act of physical love:

> Visiting churches and palaces, all of the ruins and the pillars,
> I, a responsible man, profit from making this trip.
> With my business accomplished, ah, then shall only one temple,

AMOR's temple alone, take the initiate in.
Rome, thou art a whole world, it is true, and yet without love this
World would not be the world, Rome would cease to be Rome.[29]

In the palindromic conjunction of ROMA and AMOR, the act of traveling
and the (loving) body are inseparably connected to each other.[30] Indeed, as
the following elegies clearly illustrate, the traveler is only able to artistically
portray and experience Italy so well because he does not merely *observe*
and read books, but grabs, much like Goethe's Faust, "[f]rom the whole
store of human life"[31]. Thus, the visitor to Rome admittedly avails himself
of "the works of the Ancients" (Goethe, *Elegies* VII), but also primarily of
the body of his soon found Roman lover. Rome only opens up to him in
this link between observation and physical feeling which is formulated in
the central and chiastic verses of the famous fifth elegy: I "See with an eye
that feels, feel with a hand that sees" (V).

Unlike Werther then, the traveler perceives the naked body not only in
its idealized statue-like properties and via an abstract concept of 'beauty'
as described by Winckelmann, but rather the opposite. He finally understands the marble statues of Rome at the very moment he touches the body
of his beloved and enjoys it with his senses:

Do then I not become wise when I trace with my eye her sweet bosom's
Form, and the line of her hips stroke with my hand? I acquire,
As I reflect and compare, my first understanding of marble,
See with an eye that feels, feel with a hand that sees.

(VII)

In these verses 'seeing' does not mean distant observing, but sensuous
perception, a new kind of contemplation that also leads the traveler in
Goethe's later *Italian Journey* (*Italienische Reise*, 1813–17) – which to this
day remains the most influential and popular German text of the genre – to
novel insights into a country that had already been traveled and described
by countless predecessors. In the *Roman Elegies* as well as in the *Italian
Journey*, body and mind, sense and sensuality blend harmoniously into
one and allow the traveler not only to really perceive his new surroundings
but also to produce art. Thus, the speaker of the *Elegies* admits: "In her
embrace – it's by no means unusual – I've composed poems / And the hex-
ameter's beat gently tapped out on her back, / Fingertips counting in time
with the sweet rhythmic breath of her slumber" (VII). The text is literally
inscribed on the lover's body.

In stark contrast to the first part of the *Letters from Switzerland*,
Goethe thus does not only formulate his understanding of a living classic

in the *Roman Elegies* but also marks a significant break within the genre of travel literature. For Goethe, a *Bildungsreise* no longer meant instruction through facts or rational appropriation but far more a sensual experience and its processing in poetic terms. He marked traveling as a creative act in itself, in which the physical and the aesthetic merged into one. In his emphasis on the human body, its beauty, and its sensuous qualities, he shaped and influenced the whole genre of travel literature on its way from facts to physicality,[32] and at the same time tried to accentuate the importance of a holistic understanding of body and mind, nature and art.

The latter was all the more important to him in an age that in its striving for progress and its growing reliance on science, new technical instruments and apparatuses increasingly excluded the body and its perceptional potentials from the equation. Goethe's body concept is thus also to be seen as his artistic protest against the momentous split between the two cultures around 1800 and as an urgent appeal for a human(e) science and perspective on the world.

Notes

1 Johann Wolfgang Goethe, *Fifteen Letters by Johann Wolfgang von Goethe from His First Journey to Switzerland in 1775* [*Letters From Switzerland. Part the First*], trans. Quincy Morgan Bayard (San Francisco: The Greenwood Press, 1949), p. 1. Further references to this edition are included parenthetically in the text.

2 "Die Präfiguration der Wahrnehmung und Wirklichkeitserfahrung durch Reiseführer und Reisemethodiken bestand vor allem darin, dass sie kulturelle Muster für das Reisen und für Reiseberichte etablierten und popularisierten. Diese medial verbreiteten Ratschläge und Wahrnehmungsschemata stellten Vor-Bilder sowie Wahrnehmungs- und Darstellungsmodelle bereit, die weitreichende Wirkung für die Praxis des Reisens und für die Abfassung von Reiseberichten hatten." [The prefiguration of perception and the experience of reality via travel guides and traveling methods stems from the latter's establishment and popularization of cultural patterns. These common advices and schemes of perception built pre-images and perception models, which greatly influenced the experience of travel and the writing of travelogues.] (Ansgar Nünning, "Zur mehrfachen Präfiguration/Prämediation der Wirklichkeitsdarstellung im Reisebericht: Grundzüge einer narratologischen Theorie, Typologie und Poetik der Reiseliteratur", in Marion Gymnich et. al., eds., *Points of Arrival. Travels in Time, Space, and Self. Zielpunkte, Unterwegs in Zeit, Raum und Selbst* (Tübingen: Francke, 2008), pp. 11–32 (at p. 15). Unless otherwise indicated translations of works are done by myself.

3 See below and Justin Stagl, *A History of Curiosity. The Theory of Travel 1550–1800* (London: Routledge, 1995).

4 Werther uses the term "unaussprechlich" [inexpressible] in the German original, which is significant in the context of the *Letters*, though unfortunately not adequately translatable in the English version.

5 See also Philipp Sarasin, *Reizbare Maschinen. Eine Geschichte des Körpers 1765–1914* (Frankfurt am Main: Suhrkamp, 2005).

6 Rudolf Behrens and Roland Galle, "Einleitung", in Rudolf Behrens and Roland Galle, eds., *Leib-Zeichen. Körperbilder, Rhetorik und Anthropologie im 18. Jahrhundert* (Würzburg: Königshausen und Neumann, 1993), pp. 7–9 (at p. 7).

7 Torsten Hoffmann, *Körperpoetiken. Zur Funktion des Körpers in der Dichtungstheorie des 18. Jahrhunderts* (Paderborn: Wilhelm Fink, 2015), p. 42.

8 Although presumably both written in the 1790s, Goethe published the second part of his *Letters from Switzerland* much earlier (a first selection of the letters was published in 1796 in Schiller's *Horen*) than the first, which was published in 1808.

9 See also Uwe Hentschel, "Goethe und die Reiseliteratur am Ende des Achtzehnten Jahrhunderts", in Anne Bohnenkamp, ed., *Jahrbuch des Freien Deutschen Hochstifts, 1993* (Göttingen: Wallstein Verlag, 1993), pp. 93–127.

10 "Das gemeine Absehen bey Reisen soll gemeiniglich darinnen bestehen, daß man die Welt kennen lerne, das ist, die Völcker in ihren Sitten, Gewohnheiten, Aufführungen betrachtet, und alles gehöriger massen zu seinem Nutzen anwendet." ("Reisen", *Zedlers Grosses vollständiges Universallexicon aller Wissenschafften und Künste*, vol. 31, ed. Johann Heinrich Zedler (Halle/Leipzig: Johann Heinrich Zedler, 1731–54), pp. 366–85 (at p. 366)).

11 See, for example, François Maximilien Misson, *Voyage d'Italie* (Utrecht: van de Water and van Poolsum, 1722); Joachim Christoph Nemeitz, *Nachlese besonderer Nachrichten von Italien* (Leipzig: Gleditsch, 1726); and Johann Georg Keyßler, *Neueste Reise durch Teutschland, Böhmen, Ungarn, die Schweiz, Italien und Lothringen* (Hannover: Förster, 1751).

12 Friedrich Nicolai, "Kleine Nebenreise von Wien nach der Gränze von Ungarn und nach Wien zurück", in Friedrich Nicolai, *Reise durch Deutschland* (Berlin/Stettin: Nicolai, 1785).

13 "Daß wenig Reisebeschreibungen so lehrreich, mit so feinem Beobachtungsgeiste, so edler Freymüthigkeit und so unparteyischer Wahrheitsliebe abgefaßt sind als die *Nicolaische*." ([Anonymous], "Beschreibung einer Reise durch Deutschland und die Schweiz, im Jahre 1781" [Review], *Allgemeine Literaturzeitung*, 4, no. 239 (1786): 33).

14 Heinrich Heine, *The Harz Journey and Selected Prose*, trans. and ed. Richie Robertson (New York: Penguin Classics, 2006), pp. 34–35.

15 Novalis, *Philosophical Writings*, trans. and ed. Margaret Mahony Stoljar (New York: The State University New York Press, 1997), p. 25 (emphasis in text).

16 Jürgen Helm and Karin Stukenbrock, eds., *Anatomie. Sektionen einer medizinischen Wissenschaft im 18. Jahrhundert* (Stuttgart: Franz Steiner Verlag, 2003), p. 7.

17 See also Josef N. Neumann and Rüdiger Schultka, eds., *Anatomie und Anatomische Sammlungen im 18. Jahrhundert. Anlässlich der 250. Wiederkehr des Geburtstages von Philipp Friedrich Theodor Meckel (1755–1803)* (Berlin: LIT, 2007); and Helm and Stukenbrock, *Anatomie*.

18 Hoffmann, *Körperpoetiken*, p. 10.

19 Marianne Henn and Holger A. Pausch, "Introduction: Genealogy and Construction of Body Identity in the Age of Goethe", in Marianne Henn and Holger A. Pausch, eds., *Body Dialectics in the Age of Goethe*, Amsterdamer Beiträge zur neueren Germanistik, vol. 55 (Leiden/Boston: Brill, 2003), pp. 9–21 (at p. 20).

20 See also K.F. Hilliard, "Atemübungen: Geist und Körper in der Lyrik des 18. Jahrhunderts", in Henn and Pausch, eds., *Body Dialectics*, pp. 293–313.

21 See also Robin Jarvis in his chapter "'Indolence Capable of Energies': Coleridge the Walker" (pp. 126–54), who finds similar poetic strategies in Samuel Taylor

Coleridge's writing (Robin Jarvis, *Romantic Writing and Pedestrian Travel* (London: Macmillan Press, 1997)).

22 Johann Wolfgang von Goethe, *Wilhelm Meister's Apprenticeship and Travels*, ed. Thomas Carlyle (Philadelphia: Lea and Blanchard, 1840), p. 272.

23 Goethe, *Wilhelm Meister's Apprenticeship*, p. 272.

24 See also Alfred G. Steer, "The Wound and the Physician in Goethe's *Wilhelm Meister*", in Siegfried Mews, ed., *Studies in German Literature of the Nineteenth and Twentieth Centuries. Festschrift für Frederic E. Coenen* (Chapel Hill: The University of North Carolina Press, 1970), pp. 11–23; Gloria Flaherty, "The Stage-Struck Wilhelm Meister and the 18th-Century Psychiatric Medicine", *Modern Language Notes*, 10, no. 3 (1986): 493–515; and Robert D. Tobin, *Doctor's Orders. Goethe and Enlightenment Thought* (New York et al.: Bucknell University Press, 2001).

25 Johann Wolfgang von Goethe, *Wilhelm Meister's Journeyman Years or The Renunciants*, trans. Krishna Winston, in Jane K. Brown, ed., *Goethe. The Collected Works*, vol. 10 (Princeton: Princeton University Press, 1995), p. 322.

26 Manfred Link, "Goethes Wertheriade *Briefe aus der Schweiz. Erste Abteilung*", in *Doitsu Bungaku*, 52 (1964): 107–20 (at p. 108).

27 This visit to a brothel did not only shock Goethe's contemporaries and Goethe scholars of various generations, who mostly chose to ignore the text, but was also shamefully left out in the first English translations of the *Letters*, which in its place, without any indication of the omission, featured a bridging text that was not even written by Goethe. To this day, there are only few editions that present a translation of the unabridged and original text.

28 Goethe presumably refers to Heinrich Julius von Lindau, a Hessian officer, who fought in North America and died in action in 1776.

29 Johann Wolfgang von Goethe, *Erotica Romana* (The Floating Press [e-book], 2013), VII. The Roman numeral indicates the corresponding number of the elegy. Further references to this edition are included parenthetically in the text.

30 See also Wulf Segebrecht, "Sinnliche Wahrnehmung Roms. Zu Goethes *Römischen Elegien*, unter besonderer Berücksichtigung der *Fünften Elegie*", in Wulf Segebrecht, ed., *Gedichte und Interpretationen. Bd. III: Klassik und Romantik* (Stuttgart: Reclam, 1984), pp. 49–59.

31 Johann Wolfgang von Goethe, *Faust I and II*, trans. and ed. Stuart Atkins (Princeton: Princeton University Press, 2014), p. 6.

32 Goethe's sensuous approach to travel was subsequently picked up by various writers and authors of travelogues. Especially the *Italian Journey* became a role model for German travel literature in the nineteenth century, and its influence is traceable to this day. See, for example, Durs Grünbein, *Aroma. Ein römisches Zeichenbuch* (Frankfurt am Main: Suhrkamp, 2010) and (though in a negative reaction to Goethe, turning the sensuous physicality to pornography) Rolf Dieter Brinkmann, *Rom, Blicke* (Reinbek bei Hamburg: Rowohlt, 1979).

4 Motherhood and the Embodied Traveler in Wollstonecraft's *Letters Written during a Short Residence in Sweden, Norway, and Denmark*

Elizabeth Zold

4.1 Introduction

Mary Wollstonecraft's most commercially successful text, highly praised by the reading public,[1] and the one that is often admired by biographers for its openness and primacy of feelings, *Letters Written during a Short Residence in Sweden, Norway, and Denmark* (1796), has always been marked by its deeply personal content.[2] The travelogue is a written account of the journey Wollstonecraft was sent on by her lover, Gilbert Imlay, to Scandinavia in the summer of 1795 to recover his ship full of silver that had gone missing, presumed stolen, and then sold off by its Norwegian captain, Peter Ellefsen. While the two were not legally married, Imlay had sent along a letter that designated her as "Mary Imlay my best friend and wife [sic]" (*A Short Residence* Appendix 1). Wollstonecraft traveled with one-year-old Fanny, her daughter with Imlay, and Marguerite, Fanny's nurse-maid. Wollstonecraft kept a journal and wrote letters to Imlay throughout her journey, capturing observations, personal feelings, and business details for him. She had attempted suicide mere weeks before the trip because of Imlay's infidelity, and because their relationship had ended at the time of her travels, Wollstonecraft's emotions are in the forefront of the personal letters.

Wollstonecraft wrote her travelogue after Imlay returned her letters in November of 1795,[3] and it is notable that the book was born out of devastating personal circumstances: some scholars, such as Margot Beard, argue that Wollstonecraft decided to edit and sell the letters as a travel narrative as a way to gain economic freedom for herself as a single mother.[4] Indeed, as Janet Todd notes, Wollstonecraft had no choice but to be a writer so as to "keep herself and Fanny".[5] As Wollstonecraft notes in the advertisement of her book, the letters were "designed for publication" (*A Short Residence* 3), although she declines mentioning that they had been edited greatly from the originals and the original recipient's name had been removed, with only hints of a tumultuous relationship between them. It is

DOI: 10.4324/9781003331803-5

surprising that this emotion-filled travel narrative about a mother's travels was so successful, in part because travel and motherhood were incompatible according to eighteenth-century middle-class English values. Mothers were meant to represent stability, and thus social mores expected them to remain in the home; mothers who traveled could be accused of maternal neglect. As I will discuss below, the political and ideological constructions of motherhood, especially absent mothers, in the long eighteenth century have been given more scholarly attention in the past few decades. Still, there is ample space for the examination of the role the construction of motherhood plays in eighteenth-century women's travelogues, especially for the few mothers like Lady Mary Wortley Montagu (1763)[6] and Lady Elizabeth Craven (1789),[7] who, like Wollstonecraft, brought their children along for the journey.

Within her travelogue, Wollstonecraft writes from a multitude of subjectivities: mother, slighted lover, philosopher, traveler, and feminist, and the text reflects the concerns of each of them. Given the personal nature of the letters, which are a departure from her earlier writings, *A Short Residence* has often been read as a personal journey, albeit with political observations and economic commentary, through which Wollstonecraft works to construct her own subjectivity. Perhaps owing much to William Godwin's assertion, "[i]f ever there was a book calculated to make a man in love with its author, this appears to me to be the book",[8] many scholars focus on the personal aspect of the narrative. Eleanor Ty (1992),[9] Karen Lawrence (1996),[10] Nancy Yousef (1999),[11] and Margot Beard (2004)[12] all examine the ways in which Wollstonecraft works through her subjectivity, identity, and moral philosophies in *A Short Residence*, often noting the duality of the text, which blends travelogue and autobiographical elements as well as emotion and reason.[13] Certainly, Wollstonecraft's use of sensibility is striking given her well-documented aversion to the discourse and the gendering of the sentimental novel, which Mitzi Meyers explores in her article "Sensibility and the Walk of Reason" (1990).[14] Elizabeth Bohls's *Women Travel Writers and the Language of Aesthetics 1716–1818* (1995)[15] steps away from this line of inquiry to focus on Wollstonecraft's new way of understanding the relationship between the aesthetic subject and aesthetic object and the way Wollstonecraft connects the body's sensuous pleasures to larger social processes and power relations in society. How Wollstonecraft used and shaped aesthetic discourses within *A Short Residence* coincides with motherhood in Jeanne Moskal's essay "The Picturesque and the Affectionate in Wollstonecraft's *Letters from Norway*" (1991).[16] Moskal argues that Wollstonecraft's affection for Fanny shapes her use of the picturesque and the ways in which she conceptualizes the gendered aesthetic conventions of the sublime and the beautiful.

However, very little scholarship has focused on the role of motherhood specifically in *A Short Residence*; most of the scholarly discussion surrounding Wollstonecraft and motherhood concentrates on Wollstonecraft's novel *Maria; or the Wrongs of Woman* (1798)[17] or her essays such as *Thoughts on the Education of Daughters* (1787)[18] and *Vindication of the Rights of Woman* (1792).[19] Regarding this aspect, the scholarly directions are quite varied. Thomas H. Ford's article, "Wollstonecraft and the Motherhood of Feminism" (2009) discusses the rhetoric used to talk about motherhood throughout her texts, including her fiction, essays, and *A Short Residence*. Ford claims that Wollstonecraft uses discussions between mothers and daughters paradoxically in order to suggest "an alternative and nonmaternal form of female solidarity, of a cross-generational feminist community lying beyond shared biology".[20] Moving in a radically different direction, Rachel Seiler-Smith utilizes a number of Wollstonecraft's fiction and essays to examine her use of morbid stories, including but not limited to infanticide, suicide, and rape.[21] According to Seiler-Smith, rather than showing maternity as a hopeful method of bringing new (political) life, Wollstonecraft imagines and threatens macabre ends to women and children – and, by extension, the state – if their conditions are not altered. Deviating from broader discourses like aesthetics and politics and focusing on the role of the family as a whole, Natsuko Hirakura's essay "The Portrait of a Family: Wollstonecraft's *Letters from Sweden*" (2011)[22] contends that the edited letters in *A Short Residence* show Wollstonecraft's attempt to reunite her family. Hirakura argues that by tracing Wollstonecraft's evolving persona over the course of the letters, the reason Wollstonecraft agreed to the journey in the first place becomes clear: her longing to reunite with her family.

Nevertheless, there is still space in which to examine how Wollstonecraft represents motherhood in her travelogue; as such, how Wollstonecraft situates her subject position as a mother traveling with her daughter is the specific focus of my inquiry into *A Short Residence*. I argue that, in addition to the text, establishing Wollstonecraft's commitment to her daughter, as well as her dedication to specific maternal ideals she expressed in prior publications, Fanny's presence in the travelogue goes beyond a demonstration of maternal duty. Moreover, travel writing places Wollstonecraft physically into the public sphere in a way that other types of writing such as poetry or the novel would not, since they could be written from the comfort of the domestic sphere. However, travel writing demanded literal embodiment and exposure in the public sphere; as a result, travel writing requires a corporeal component for its creation. I argue that Wollstonecraft, embodying motherhood, demonstrates the important connection between the mind and body for authentic travel writing. Wollstonecraft first draws attention to the traditional erasure of the embodied observer within travel literature by the very act of being a mother: the reader assumes that she is not, by virtue of reproducing, disembodied. She seeks to acknowledge this bodily element, using Fanny's presence in the text as a locus for Wollstonecraft to explore the relationship between emotion and reason, and a way for her to demonstrate how the feeling body enhances rather than detracts from the authenticity of a travel narrative. That is, Wollstonecraft demonstrates via her maternal body that perception depends upon individuals and their understanding of interactions with the world, mitigated through their individual bodies. As such, her narrative argues that authentic travel writing would be a truthful account that encompasses the author's feelings and understanding of a moment: authenticity requires that travel narratives necessarily cannot be completely objective, as experiences will always be filtered through the individual.

4.2   The Distancing 'Eye' and the Invisible 'I': Eighteenth-Century Travelogue Conventions

By the eighteenth century, the travelogue genre was governed by and steeped in textual conventions that, when followed correctly by authors, were signs of an authentic travel narrative; that is, the text was considered truthful and accurate, written by someone who had actually traveled, writing down what they saw and experienced. Carl Thompson notes in his overview of the travel writing genre that "if travelogues are to be credited by their readers, they must meet contemporary audience expectations as to what denotes reliability and plausibility in the travel account".[23] Michael McKeon explains in *The Origins of the English Novel, 1600–1740* (1987) that early-seventeenth-century publications of the Royal Society and their instructions for travel writers were the formative sources of early travelogue guidelines, which promoted geographical and anthropological notes written in an objective style as a way to help prove the truthfulness, and thus authenticity, of the narrative.[24] Jean Viviès agrees, arguing that the Royal Society influenced the travelogue genre's "tacit if not written rules of conduct".[25] Of course, these conventions evolved over the course of the eighteenth century. Charles Batten, in his formalist analysis of the conventions of eighteenth-century travel literature, lays out their development over the course of the century. According to Batten, travelers knew not to talk about themselves in order to avoid charges of egotism, to write in a plain style and avoid the ornaments of rhetoric, and to instead make use of encyclopedic categorization of what they saw, by writing on the spot instead of waiting and later trying to remember.[26] Avoiding egotism meant keeping the traveler, as well as his or her opinions and feelings, out of the text. A genre that relied on conventions to substantiate objectivity and truth, the narrative of travel eschewed narrators and their personal, embodied experiences, instead focusing outward on new people and places encountered.

Within the later decades of the century and continuing well into the nineteenth century, critical discourse on the travelogue's function centered on the tension between objective and subjective narration, with the more objective, 'scientific' discourse taking precedence and authority over more personal travel narratives. While the travelogue genre would become more subjective and written for pleasure during the nineteenth century, with a focus on the Romantic subject's connection with nature, this shift did not end the debate over the style and nature of the travelogue. Nigel Leask argues that much scholarship on Romantic travel literature has focused too closely on the personal narratives and maintains that by the mid-eighteenth century through the 1820s, "the rise of scientific disciplinarity [...] betokened the need to evacuate the anecdotal, interactive 'personal narratives' of travel in order to establish a more objective epistemological authority".[27] Indeed, Samuel Taylor Coleridge, in a *Table Talk* entry from 1833, states that he "wish[es] the naval and military officers who write accounts of their travels would just spare us the sentiment. The Magazines introduced this cant. Let these gentlemen read and imitate the old captains and admirals, as, Dampier &c".[28] While subjective narratives may have gained more popularity in the latter half of the century, narratives that included objective, scientific discourse were still considered more truthful, and thus more authoritative.

In fact, eighteenth-century epistemological discourses often used travel literature to create what were considered authentic texts, removing the acknowledgment of the feeling body. Joseph Addison, in his essay "The Pleasures of the Imagination" in the *Spectator* no. 411 (1712), declares: "Our Sight is the most perfect and most delightful of all our senses [...] and may be considered as a more delicate and diffusive kind of touch, that spreads itself, over an infinite multitude of bodies, comprehends the largest Figures, and brings into our reach some of the most remote parts of the universe".[29] Addison, like others, argued for the supremacy of sight for understanding the world, believing that sight allowed the observer/explorer to take in massive amounts of information at once, certainly more than the other senses. However, he cannot escape from using the language of touch to describe the way in which sight works.

Discussing sight in terms of touch becomes an even more interesting metaphor as the physicality of touch was eschewed in favor of sight, a mode that allows for more distance between the subject and the object. The image of the explorer's vision "that spreads itself over an infinite multitude of Bodies" played a large role in what would become England's empire and was what Mary Louise Pratt describes as the imperial eyes of European colonizers in *Imperial Eyes: Travel Writing and Transculturation* (1992).

Pratt argues that through travelogue accounts that describe new countries as presenting themselves to travelers, "the landscanning European eye seems powerless to act upon or interact with this landscape that offers itself"[30] with all of the descriptions "emanating from a seat of power behind the invisible, innocent speaking 'I'".[31] In disembodying the physical subject of the 'I' behind the 'eye' travelers maintained a distance between themselves and the observed objects. Matthew Edney notes that in the eighteenth century, sight was "the most mechanic, least sensuous"[32] sense and thus, more closely related to the truth. Rooted in vision, eighteenth-century epistemology, then, "established almost a physical distance between the viewer and viewed, between the subject and object of vision".[33] The body became distanced from experience, and as a result, the erasure of the embodied self eliminated the individuality of the writer. That is, if the 'eye' is objective and simply records what it sees to exist in the world, the knowledge gained can be transferred to any person or group.

As a result, Erin Goss argues in *Revealing Bodies* (2013), the body is continually disavowed in the long eighteenth century, and even within contemporary discourses, because including it would disrupt the knowledge that depends on generality.[34] That is, the body/a body and its role within the construction of knowledge are complex because generalized knowledge requires a generalized conception of the body; thus, the notion that each person has an individual, specific body is often removed. That is to say, as Goss meticulously examines, empiricism assumes that bodies are all similar in form and structure, as well as in minute details and sensory organs. The idea that every single body is unique threatens the notion that people can know and create generalized knowledge through experience because each particular person would necessarily have a slightly different experience.[35] As a result, the erasure of the embodied self eliminates the individuality of the writer.

However, as Wollstonecraft illustrates throughout her narrative, travelers have feelings, emotions, and experiences that individualize and form how they interpret their experiences, making them specific to each individual. In *A Short Residence*, Wollstonecraft brings together objective descriptions and personal feelings, blending together the narrative styles of the detached travel writer and the embodied, feeling traveler. In fact, she eschews the convention of not talking about oneself, which she explicitly states in the ad for *A Short Residence*, confessing, "I found I could not avoid being continually the first person—'the little hero of each tale'" (3). As Caroline Franklin remarks, Wollstonecraft put "a much greater emphasis than had ever been seen before in travel writing on scenery description and the author's lyrical response to the sublime" but does not completely reject the notion of a traveler as a gatherer of facts and knowledge.[36] Within this blending of narrative styles, Wollstonecraft makes the 'I' behind the observing 'eye' explicit in part through her embodied position as a mother, and thus, Wollstonecraft illuminates how the self and its effect on the text had been ignored in the majority of purported 'objective' travel literature. As I will demonstrate, Fanny is often set at odds with her mother's experiences in the text, living a blissfully innocent life of a child, highlighting the subjective nature of experience. Wollstonecraft demonstrates how, since information learned through observation is filtered through the senses, the thinking, feeling self is needed to represent the fullness of the travel experience for the reader.

4.3   Mother and Daughter: Making the 'I' Visible

Wollstonecraft uses her feelings to move into musings about anxieties concerning the struggles Fanny will face as a woman. It is not Wollstonecraft as a mother who will do harm to Fanny, it is society at large. She writes: "You know that as a female I am particularly attached to her—I feel more than a mother's fondness and anxiety, when I reflect on the dependent and oppressed state of her sex" (*A Short Residence* 36). No longer seeing her child as an extension of herself, Wollstonecraft steps back to view Fanny as an independent person. She carefully situates her concern not within the rhetoric of the overly emotional mother, but as a reasoning woman who is aware of the pains and oppression another woman will suffer. Showing how much she cares for her daughter, yet remaining within the confines of rational affection, Wollstonecraft disavows herself as a mother so as to be concerned about a fellow woman. But then, Wollstonecraft pulls back from her social critique and states "[b]ut whither am I wandering?", before going back to her observations about the kindness of the Norwegians and their customs for how they treat their houseguests (36). The incongruity of switching from the heavy reflection upon her daughter's future oppression to that of more objective observations regarding the hospitality customs of her host and hostess lays bare the erroneousness of believing that objective observation leads to authentic narration. Wollstonecraft's discussion of her fears is a piece of her experience in that moment, embodying not just motherhood but her gender more broadly. Her embodied identities inform how she understands other cultures because of her lived experiences as a woman: her body has determined her interactions with the world.

Within *A Short Residence*, Wollstonecraft presents her reasoned musings on gender as stemming from her emotions for Fanny, arguing that the relationship between reason and feeling is one of cause and effect within the context of motherhood. She then transfers that same idea to emotions and reason more generally, arguing that the oppression of women has given them cause to think deeply because of the intense emotions they have felt. In relaying an anecdote about the commonality of Danish middle-class men having affairs with their servants, Wollstonecraft takes a paragraph to lament gendered power dynamics across Europe. Then she pulls back and notes: "Still harping on the same subject, you will exclaim—How can I avoid it, when most of the struggles of an eventful life have been occasioned by the oppressed state of my sex: we reason deeply, when we forcibly feel" (107). In a powerful rhetorical move, rather than disavow emotion altogether, Wollstonecraft argues it serves as an *impetus* for thought; feelings become a part of the process of reason, rather than a hindrance to it. Rather abruptly, then, Wollstonecraft writes: "But to return to the straight road of observation" (107). Her sudden turn away from her impassioned thoughts about the systemic oppression of women in order to simply 'observe' what is around her underscores the inadequacy of the travelogue convention of disembodied, impersonal narration to nurture an authentic narrative. Instead, her passion concerning the plight of her gender demonstrates how Wollstonecraft's embodied identity directly impacts her reasoning, and, as a result, how she experiences and engages with what is happening around her. That is, Wollstonecraft's authentic travel narrative – a truthful record of her experience – is necessarily informed by her lived experience as a women, and ignoring those ruminations for a listing of observations would actually lessen the authenticity of her experience in that moment. Because emotion and reason are connected for Wollstonecraft, an embodied woman and mother, she cannot help but view these reoccurring social issues from a subjective perspective. In moving away from discussing broader social issues that impact her future as well as her daughter's simply to keep to a timeworn convention that gives an air of objectivity and reason, Wollstonecraft points out the easily performative nature of a convention meant to mark authenticity.

But Fanny's presence does not just lead to musings on significant social issues: throughout the text, she is used as a contrast for what Wollstonecraft experiences as a traveler. In comparing Fanny's travel experience to her own, Wollstonecraft reiterates how each traveler's emotions affect how they interpret and experience an object based on their education and life experience. Early on in the text, Wollstonecraft's eyes catch sight of some "heart's-ease", which she takes as a good omen and "going to preserve it in a letter that had not conveyed balm to my heart, a cruel remembrance suffused my eyes; but it passed away like an April shower" (9). Such a flash of mixed emotions – no doubt influenced by her complicated relationship with Imlay – is followed by a more intellectual and academic approach to describing the flower. Wollstonecraft reminds her recipient that if they were "a deep read in Shakespeare", they would know heart's-ease is a little flower "tinged by love's dart" (9) as described in *A Midsummer Night's Dream*.[37] Wollstonecraft's "mental obstacle course to find meaning in the land"[38] is offset against her daughter's reaction, as Wollstonecraft points out that "[t]he gaiety of my babe was unmixed; regardless of omens or sentiments, she found a few wild strawberries more grateful than flowers or fancies" (*A Short Residence* 9). While her mother's reaction to the surroundings is an emotional one that leads her to reflect on the more erudite connotations of the flower, Fanny appreciates the simpler, gustatory pleasures of the journey. Unburdened by either the knowledge of the flower's literary meaning or the feelings it produces about love in her mother, Fanny's emotional relationship to the flower within the exact time and place is one of disinterest, as it is less appealing than satisfying her immediate bodily senses. Set against her daughter's interaction with the surroundings, Wollstonecraft's description of her reaction to the flower lends authenticity precisely through its expression of individuality.

Fanny continues to act as a contrast for her mother, with Wollstonecraft utilizing spatial metaphors to articulate the vastly different experiences and emotions the journey arouses in the two. While awake one evening, Wollstonecraft contemplates the rocky landscape, pondering what could be keeping her from sleep, writing, "[w]hy fly my thoughts abroad when every thing around me appears at home? My child was sleeping with equal calmness—innocent and sweet as the closing flowers" (11). Wollstonecraft views Fanny as mirroring the flora around her, blissfully unaware of her parent's relationship strife while her mother's thoughts wander with the late evening light. In being compared to flowers, Fanny is connected to the earth, metaphorically grounded while her mother's thoughts soar in the sky. Fanny also makes herself at home while her mother's thoughts "fly abroad": although they are traveling, Fanny is seemingly content in their new surroundings, and in doing so, becomes a source of comfort for her mother. She is the embodiment of home, a familiar constant during their travels. Wollstonecraft admits that "[s]ome recollections, attached to the idea of home" intermixed with "reflections respecting the state of society [that she had] been contemplating that evening" leads her to shed a tear, which falls upon Fanny's cheek (11). Wollstonecraft's tears are a physical manifestation of her emotion, a reminder of the embodied, feeling traveler. As a mother traveling and sharing a quiet moment with her daughter, Wollstonecraft exhibits how larger life experiences and knowledge necessarily influence how a traveler experiences and understands an event.

4.4 Motherhood and Travel in the Eighteenth Century

Discussing her daughter in her published travel narrative, even if it does show Wollstonecraft's maternal side, was a risky endeavor given that motherhood was intensely scrutinized over the course of the eighteenth century. The rise of middle-class values in the eighteenth century gave way to a new formation of the domestic sphere that placed the mother at the center of the moral, familial realm. Scholars such as Nancy Armstrong (1987),[39] Toni Bowers (1996),[40] Katherine Turner (2001),[41] and Marilyn Francus

(2012)[42] have argued that eighteenth-century conduct manuals and novels played a large role in constructing the discourse surrounding domesticity, motherhood, and maternity. Specifically, much of the literature argued the idea of the 'natural' duty of women to be in the home. In *The Politics of Motherhood*, Bowers locates the rise of the ideal domestic woman early in the century and extends Armstrong's discussion of the creation of the ideal eighteenth-century domestic woman by emphasizing the idea of motherhood as a woman's 'natural' duty. Elizabeth Johnston explains that conduct books worked to create a simplified, homogeneous narrative of mothering: one of 'natural' selflessness and nurturing.[43] However, there were limits and rules placed upon nurturing as well. As Julie Kipp writes, Romantic motherhood encouraged women to indulge their excessive emotions to bond with their children, but at the same time, mothers were told to also channel and tame that sentimentality for the interests of the child, since overindulging could hurt the child and eventually, the nation.[44] As the key figures in rearing the future generation of citizens, mothers were required to walk a very thin line between love and overindulgence, quite often with little guidance as to the difference between them.

One particularly divisive issue in regard to mother and child bonding was breastfeeding. Upper-class mothers could afford to hire wet nurses and maids, and often did, especially early in the century;[45] as a result, Augustan conduct manuals often vilified aristocratic women as lacking in maternal feelings, arguing that they saw breastfeeding as an inconvenience.[46] The pressure on mothers across social classes to breastfeed and maintain close physical contact as a way of showing maternal affection continued through the end of the century, with Wollstonecraft herself entering the debate in her conduct book, *Thoughts on the Education of Daughters* (1787).

Within the discourse on maternal feelings and breastfeeding, Wollstonecraft argued for a more nuanced understanding of how women bonded with their children through nursing. She challenged the notion that women could better bond with children because they 'naturally' had more emotions than men. Instead, in *Thoughts on the Education of Daughters*, Wollstonecraft advocates for breastfeeding as a way to produce maternal affection. She argues that the helplessness and dependence of the baby on the mother, as well as the physical closeness, will bring forth maternal feelings.[47] Wollstonecraft claims that it is through maternal bonding that a woman produces maternal love and what she terms "rational affection" (5). That is, the sentiments are elicited in women through their maternal actions, and the emotions themselves can be deemed rational, not excessive. She complicates the idea of maternal instinct and eschews the notion of women as having more innate feelings, instead arguing that maternal feelings are formed "quite as much from habit as instinct" (4). According to Wollstonecraft, it is the maternal actions that form the feelings, not the other way around: therefore, she instructs women that each should perform her duty as a mother in order to feel the "rational affection for her offspring" (5).

Wollstonecraft's viewpoint on breastfeeding reveals itself during her travels. She mentions that she had nursed Fanny herself, despite the physical difficulty it brought her. In *A Short Residence*, she writes that some "imprudence" and "untoward accidents" occurred while she was weaning Fanny earlier that year which had "reduced me to a state of weakness which I never before experienced" (50). In an interesting rhetorical move, Wollstonecraft does not place the blame for her prior physical weakness on nursing her daughter. She tempers its physical impact on her by alluding to other life events that coincided with the end of her time breastfeeding Fanny.[48] Wollstonecraft's connection to her daughter may have been strengthened through the physical act of nursing, but she is clear that it required a physical sacrifice on her part. Of course, the fact that she was weaned several months prior to travel is what gives Wollstonecraft the freedom to travel, as weaning allows for a literal physical distance between mother and child. Interestingly, while motherhood weakened her body, Wollstonecraft gives credit to her current travels for renewing her physical strength. She writes that her "constitution has been renovated here; and that I have recovered my activity, even whilst attaining a little *embonpoint*" (50, emphasis in text). Thus, rather than neglecting her maternal duties during her journey, Wollstonecraft's health is improved, the evidence of which manifests itself physically, an even more important development as Wollstonecraft will end up raising Fanny alone.

In contrast to Wollstonecraft fulfilling her maternal duty to physically nurture her daughter, Swedish upper-class women used wet nurses, who were primarily women from the lower classes. Wollstonecraft critiques this choice in her travel narrative, noting that "the total want of chastity in the lower class of women frequently renders them very unfit for the trust" (23). Concentrating on the lack of "chastity" of the nursing bodies, Wollstonecraft translates this perceived moral failing as a corruption of their bodies, which renders them unsuitable for nourishing a child. Such is the danger of commodifying such a pivotal child-rearing duty: it reduces nursing to bodily labor rather than also doing the work of strengthening familial bonds. But the detrimental effect of treating maternal responsibilities as commerce affects the family unit as a whole, not just mother and child, as Wollstonecraft observes in Norway. She relays an anecdote about the wet nurse of the woman in whose inn she is residing. Wollstonecraft notes that the young woman is paid 12 dollars a year, but in return must pay her own wet nurse ten dollars a year. As a result, "the father had run away to get clear of the expence [sic]" (54). The domestic situation that so closely mirrors her own draws her compassion as she ponders how quickly happiness can be shattered. "It was too early for thee to be abandoned, thought I" (54), she writes, in a double meaning that would become achingly apparent. In such a situation, monetizing such an integral maternal duty led to the breakup of the family unit.

4.5   'Rational Affection': Motherhood, Emotion, and Reason

The travelogue is expurgated of the majority of emotions and personal details Wollstonecraft discussed in her original letters to Imlay, presumably to make the letters more suitable for publication as a travel narrative.[49] In those letters, Fanny acts as the bonding agent between Wollstonecraft and Imlay; she is the only connection Wollstonecraft still has to her distant ex-lover. While it is not within the parameters of this study to compare and contrast the two sets of letters closely, it is pertinent to mention that Wollstonecraft removes specific maternal worries about Fanny's health and safety for *A Short Residence*.[50] Fanny was teething at the time of travel, a pain that Wollstonecraft mentions a number of times in her private letters but does not mention in the travelogue. Furthermore, she confesses to Imlay, upon leaving her daughter and Marguerite, the nanny, behind in Gothenburg, Sweden while she went on to Tønsberg, Norway, "I felt more at leaving my child, than I thought I should—and [...] I asked myself how I could think of parting with her ever, of leaving her thus helpless?"[51] While these worries and thoughts of a mother leaving a child for the first time can be expected, such a statement manages to invoke ideas of both excessive maternal emotions and maternal neglect. The travel narrative instead presents a balance between reason and emotion, no doubt to contradict those who found women, especially mothers, to be overly emotional.

Given the primacy of physical proximity between mother and child, Wollstonecraft no doubt felt the need to deter any critiques, especially when the trip to Tønsberg stretched to triple the length from the anticipated one week. The tone in the travelogue is carefully poised – employing "rational affection" (*A Short Residence* 5) to use Wollstonecraft's term in her conduct book – between missing her daughter while avoiding the realm of sentimentality. When she first mentions Fanny's absence, it is a brief aside to explain her small traveling party: "This was all the party; for, not intending to make a long stay, I left my little girl behind me" (25). When Wollstonecraft arrives in Tønsberg and learns of the trip's extension, she briefly laments not bringing Fanny with her (38).

However, as her solo journey wears on and the time grows closer for her trip back to Gothenburg, Wollstonecraft allows sentiment into the discussion of her time away from Fanny:

At Gothenburg I shall embrace my *Fannikin*, probably she will not
know me again—and I shall be hurt if she do not [sic]. How childish
is this! still, it is a natural feeling. I would not permit myself to indulge
the 'thick coming fears' of fondness, whilst I was detained by business.

(73, emphasis in text)

Dubbing Fanny with a pet name adds to the sentiment of the passage as
Wollstonecraft confesses her excitement and apprehension at seeing her
daughter once again: in acknowledging her feelings, Wollstonecraft takes
control of them, allowing her to demonstrate her maternal affection while
at the same time suppressing them. Balancing commerce with motherhood,
Wollstonecraft pushes aside her maternal fears in order to conduct the
business at hand. As such, she presents herself as a mother with enough
reason to be able to control her emotions rather than letting them control
her, while at the same time admitting her deepest fears about the repercus-
sions of spending time away from her child. As Moskal notes, "throughout
the actual travel, then, one might anticipate that the role of mother grew
in importance to her as she faced the loss of the man she considered her
husband and retained only her daughter".[52]

Indeed, these fears seem to fuel her determination to be reunited with
her daughter as quickly as possible, perhaps even more so since she was
unsuccessful in locating Imlay's ship, calling the whole trip "a wild-goose
chace [sic]" (*A Short Residence* 74).[53] Wollstonecraft's tone remains matter
of fact until her urgency to reunite her little family begins to seep into the
letters. Unfortunately, upon leaving Fredrikstad in the afternoon in order to
reach Strömstad before nightfall so she could order horses for the following day "without having any thing to detain me from my little girl" (91),
her journey is hindered by a lost pilot. Thence, her tight travel schedule is
thrown off, setting forth a chain of unfortunate encounters with late car-
riages, stubborn postilions, and filthy inns. Wollstonecraft writes, "I was
particularly impatient at the last post, as I longed to assure myself that my
child was well" (94). She remains calm but impatient – reasonably affectionate in wanting to see her daughter – until she comes across a quaint tableau
of a family. A young girl is riding a horse, with her father walking next to
her holding another child while yet another young child walks behind the
cart full of rye as the four of them make their way home to the cottage where
the mother is cooking, awaiting their arrival. The perfect picture of a family strikes at Wollstonecraft's heart: "I was returning to my babe, who may
never experience a father's care or tenderness. The bosom that nurtured her,
heaved with a pang at the thought which only an unhappy mother could
feel" (95). Recalling once again the maternal bond strengthened through
nursing, Wollstonecraft uses her position as a 'good' mother to allow her
sentimentality to emerge. As Todd remarks, Wollstonecraft "could translate

suffering into care for a child and lament for a lost domestic paradise while just hinting at sexual abandonment".[54] As she faces the prospect of single motherhood, Wollstonecraft reiterates to readers that while she left Fanny for a short time, unlike Fanny's father, she would always return.

4.6   Conclusion

Indeed, although she brought Fanny along for the journey, the loneliness of single motherhood works to underscore the solitary experience of a traveler. Toward the end of her journey, on the way back to London, she writes: "I had often endeavoured to rouse myself to observation by reflecting that I was passing through scenes which I should probably never see again, and consequently ought not to omit observing; still I fell to reveries" (*A Short Residence* 116–17). Rather than giving in to conventional markers of authenticity within the travelogue genre, such as observations of what she sees, Wollstonecraft ends her last letter with the image that there is an 'I' behind the eye of observation, reminding readers that embodied, feeling travelers shape their experiences and their understanding of the experience, thereby affecting what is written in the texts. And just several lines later Wollstonecraft writes of the solitary experience of the traveler, despite having been reunited with Fanny and her nurse in Gothenburg: "[they] often fell asleep; and when they were awake, I might still reckon myself alone, as our train of thoughts had nothing in common" (117). The maternal bond she has with Fanny, while important, simply cannot supply the same kind of companionship as an adult partnership. Furthermore, as Wollstonecraft reminds us, there is no universal, authoritative travel experience.

A Short Residence is a deeply personal text; even if a reader does not know the extenuating personal circumstances, Wollstonecraft's prose marries together the genres of travelogue and personal letters. While *A Short Residence* is certainly an important inclusion to the body of Romantic travel writing, it is important to remember that the text was born out of material, maternal circumstances, the effects of which can be felt on every page. Within the text, Wollstonecraft constructs herself as an affectionate, rational, mother who – over the course of the journey – comes to the tumultuous realization that she will be parenting her daughter alone. But Fanny is not just a locus for emotions: Wollstonecraft positions herself in contrast to her daughter, highlighting the diversity of emotions and experiences different travelers have even when occupying the same space. Reminding readers of her embodied experiences as a woman and mother, Wollstonecraft disrupts the notion that travel can create a singular moment of general knowledge. In particular, her feelings as a mother traveling with her daughter necessarily frame her understanding of each of her experiences abroad, and in doing so, Wollstonecraft interrupts notions that the disembodied, distanced observer provides more of an authentic narrative simply by virtue of only describing what is seen.

Notes

1 Mary Favret, "*Letters Written during a Short Residence in Sweden, Norway, and Denmark*: Traveling with Mary Wollstonecraft", in Claudia L. Johnson, ed., *The Cambridge Companion to Mary Wollstonecraft* (Cambridge: Cambridge University Press, 2002), pp. 209–27 (at p. 225).

2 Mary Wollstonecraft, *Letters Written during a Short Residence in Sweden, Norway, and Denmark* (Oxford: Oxford University Press, 2009). Hereafter simply referred to as *A Short Residence*. Further references to this edition are included parenthetically in the text.

3 Jeanne Moskal, "The Picturesque and the Affectionate in Wollstonecraft's *Letters from Norway*", *Modern Language Quarterly*, 52, no. 3 (1991): 263–94 (at p. 265).

4 Margot Beard "'Whither am I Wandering?' A Journey into the Self: Mary Wollstonecraft's Travels in Scandinavia, 1795", *Literator*, 25, no. 1 (2004): 73–89 (at pp. 86–87).

5 Janet Todd, *Mary Wollstonecraft: A Revolutionary Life* (New York: Columbia University Press, 2000), p. 340.

6 Mary Wortley Montagu, *Letters of the Right Honourable Lady M—y W——y M——e*, vols. 1–3 (London: T. Becket and P. A. De Hondt, 1763).

7 Elizabeth Craven, *A Journey through the Crimea to Constantinople* (London: G.G.J. and J. Rosenson, 1789).

8 William Godwin, *Memoirs of the Author of* A Vindication of the Rights of Woman [1798] (Peterborough: Broadview Press, 2001), p. 95.

9 Eleanor Ty, "Writing as a Daughter: Autobiography in Wollstonecraft's Travelogue", in Marlene Kadar, ed., *Essays on Life Writing: From Genre to Critical Practice* (Toronto: University of Toronto Press, 1992), pp. 61–77.

10 Karen Lawrence, *Penelope Voyages: Women and Travel in the British Literary Tradition* (Ithaca: Cornell University Press, 1994).

11 Nancy Yousef, "Wollstonecraft, Rousseau and the Revision of Romantic Subjectivity", *Studies in Romanticism*, 38 (1999): 537–57.

12 Beard, "'Whither am I Wandering?'", pp. 73–89.

13 There are some critics who mistakenly read these personal letters as the actual letters to Imlay. Mary Poovey, *The Proper Lady and the Woman Writer: Ideology as Style in the Works of Mary Wollstonecraft, Mary Shelley, and Jane Austen* (Chicago: University of Chicago Press, 1984); J.G. Barker-Benfield, *The Culture of Sensibility: Sex and Society in Eighteenth–Century Britain* (Chicago: University of Chicago Press, 1992); Syndy Conger, *Mary Wollstonecraft and the Language of Sensibility* (New Jersey: Fairleigh Dickenson University Press, 1994); and Erinç Özdemir, "Hidden Polemic in Wollstonecraft's *Letters from Norway*: A Bakhtinian Reading", *Studies in Romanticism*, 47 (2008): 321–49, all conflate the two sets of letters, thus reading *A Short Residence* as a straightforward autobiographical text. One of the issues with this is that *A Short Residence* includes more of Wollstonecraft's political and economic commentary that is found in her journal than her private letters to Imlay; for more discussion of this issue, see Favret, "*Letters*".

14 Mitzi Meyers, "Sensibility and the 'Walk of Reason': Mary Wollstonecraft's Literary Reviews as Cultural Critique", in Syndy McMillen Conger, ed., *Sensibility in Transformation: Creative Resistance to Sentiment from the Augustans to the Romantics* (London: Associated University Presses, 1990), pp. 120–46.

15 Elizabeth Bohls, *Women Travel Writers and the Language of Aesthetics, 1716–1818* (Cambridge: Cambridge University Press, 1995).

16 Moskal, "The Picturesque and the Affectionate", p. 264.

17 Mary Wollstonecraft, *Maria: Or, the Wrongs of Woman* [1798] (New York: Norton, 1975).

18 Mary Wollstonecraft, *Thoughts on the Education of Daughters* (London: J. Johnson, 1787). Further references to this edition are included parenthetically in the text.

19 Mary Wollstonecraft, *Vindication of the Rights of Woman* (London: J. Johnson, 1792).

20 Thomas H. Ford, "Wollstonecraft and the Motherhood of Feminism", *Women's Studies Quarterly*, 37, no. 3/4 (2009): 189–205 (at p. 191).

21 Rachel Seiler–Smith, "Bearing/Barren Life: The Conditions of Wollstonecraft's Morbid Maternity", *European Romantic Review*, 28, no. 2 (2017): 163–83 (at p. 164).

22 Natsuko Hirakura, "The Portrait of a Family: Wollstonecraft's *Letters from Sweden*", in Michael Meyer, ed., *Romantic Explorations: Selected Papers from the Koblenz Conference of the German Society for English Romanticism* (Trier: Wissenschaftlicher Verlag Trier, 2011), pp. 229–38 (at p. 229).

23 Carl Thompson, *Travel Writing: The New Critical Idiom* (New York: Routledge, 2011), p. 72.

24 Michael McKeon, *The Origins of the English Novel, 1600–1740* (Baltimore: Johns Hopkins University Press, 1987), p. 110.

25 Jean Viviès, *English Travel Narratives in the Eighteenth Century: Exploring Genres*, trans. Claire Davison (London: Routledge, 2002), p. 33.

26 Charles Batten, *Pleasurable Instruction: Form and Convention in Eighteenth–Century Travel Literature* (Berkeley: University of California Press, 1978), pp. 58–69.

27 Nigel Leask, *Curiosity and the Aesthetics of Travel Writing 1770–1840* (Oxford: Oxford University Press, 2002), p. 6.

28 Samuel Taylor Coleridge, *Specimens of the Table Talk of the Late Samuel Taylor Coleridge*, vol. 1 (Ghent University: Harper and Bros, 1835), p. 136.

29 Joseph Addison, "The Spectator No. 411", in *The Works of the Late Right Honorable Joseph Addison, Esq.* vol. 3 (London: John Baskerville at Shakespeare's Head in the Strand, 1761), pp. 454–56 (at p. 454).

30 Mary Louise Pratt, *Imperial Eyes: Travel Writing and Transculturation* (London: Routledge, 1992), p. 61.

31 Pratt, *Imperial Eyes*, p. 62.

32 Matthew Edney, *Mapping an Empire: The Geographical Construction of British India, 1765–1843* (Chicago: University of Chicago Press, 1999), p. 48.

33 Edney, *Mapping an Empire*, p. 48.

34 Erin Goss, *Revealing Bodies: Anatomy, Allegory, and the Grounds of Knowledge in the Long Eighteenth Century* (Lewisburg: Bucknell University Press, 2013).

35 Goss, *Revealing Bodies*, p. 26.

36 Caroline Franklin, *Mary Wollstonecraft: A Literary Life* (New York: Palgrave, 2004), pp. 153–54.

37 William Shakespeare, *A Midsummer Night's Dream*, 2, scene 1 (1600): 161–80, https://shakespeare.folger.edu/shakespeares-works/a-midsummer-nights-dream/act-2-scene-1/ [14 October 2022].

38 Karen Hurst, "Facing the Maternal Sublime: Mary Wollstonecraft in Sweden", in Anka Ryall and Catherine Sandbach-Dahlström, eds., *Mary Wollstonecraft's*

Journey to Scandinavia: Essays (Stockholm: Almqvist and Wiksell International, 2003), pp. 139–63 (at p. 151).

39 Nancy Armstrong, *Desire and Domestic Fiction: A Political History of the Novel* (Oxford: Oxford University Press, 1987).

40 Toni Bowers, *The Politics of Motherhood: British Writing and Culture 1680–1760* (Cambridge: Cambridge University Press, 1996).

41 Katherine Turner, *British Travel Writers in Europe, 1750–1800: Authorship, Gender, and National Identity* (Aldershot: Ashgate, 2001).

42 Marilyn Francus, *Monstrous Motherhood: Eighteenth-Century Culture and the Ideology of Domesticity* (Baltimore: Johns Hopkins University Press, 2012).

43 Elizabeth Johnston, "Looking into the Mirror, Inscribing the Blank Slate: Eighteenth–Century Women Write about Mothering", in Catalina Florina Florescu, ed., *Disjointed Perspectives on Motherhood* (Lanham: Lexington Books, 2013), pp. 185–200 (at p. 195).

44 Julie Kipp, *Romanticism, Maternity, and the Body Politic* (Cambridge: Cambridge University Press, 2003), p. 18.

45 Francus, *Monstrous Motherhood*, p. 12.

46 Bowers, *The Politics of Motherhood*, p. 141.

47 Interestingly, breastfeeding is one of the few areas where Wollstonecraft and Jean-Jacques Rousseau agree in terms of the raising and educating of children. See Jean-Jacques Rousseau *Emilius; Or, A Treatise of Education*, vol. 1 (London: Dickson & Elliot, 1773) at p. 22, where he advocates for breastfeeding, arguing that when mothers nurse their own children, there will be a domino effect throughout society as overall social morals will be reformed and natural affections will be reestablished.

48 In a private letter to Imlay, published posthumously in *Posthumous Works of the Author of a Vindication of the Rights of Woman*, vol. 3, ed. William Godwin (London: J. Johnson, 1798), Wollstonecraft places the blame of her earlier weakness squarely on her breastfeeding: "I have entirely recovered the strength and activity I lost during the time of my nursing" (p. 177).

49 Wollstonecraft read and reviewed a great number of travelogues during her time as a critic for the *Analytical Review* between 1788 and 1797. See the scholarship of Ralph Wardle, "Mary Wollstonecraft, *Analytical Reviewer*", *PMLA*, 62, no. 4 (1947): 1000–09; Derek Roper, "Mary Wollstonecraft's Reviews", *Notes and Queries*, 5 (1958): 37–38; and Sally Stewart, "Mary Wollstonecraft's Contributions to the *Analytical Review*", *Essays in Literature*, 11, no. 2 (1984): 187–99. As a result, she knew the generic conventions and expectations well. Moreover, given that she discusses her heartbreak in the letters to Imlay, the letters were edited no doubt out of concerns for privacy. See also Michael Meyer's contribution in this volume: *Mary Wollstonecraft and the Body of her Letters, or: the Traveler Lost and Found in Scandinavia.*

50 For a more comprehensive comparison between and analysis of the two sets of letters, see Beard, "Whither am I Wandering?"

51 Godwin, *Posthumous Works*, p. 169.

52 Moskal, "The Picturesque", p. 265.

53 While the details of Imlay's business are not a part of the travelogue, a letter dated 5 September 1795 from Wollstonecraft to the Danish Prime Minister Count Bernstorff asks him to look into the matter for her (*A Short Residence* Appendix 2).

54 Todd, *Mary Wollstonecraft*, p. 368.

II
Other Bodies

5 Beasts on Board

Traveling Animals and Pacific Voyages in the First Two Ages of Exploration

Mira Shah

5.1 Introduction: Fresh Meat

When European explorers ventured to the Pacific, their ships carried all kinds of bodies: from the first circumnavigational voyages in the First Age of Exploration onwards, European maritime enterprises drew their crew from a diverse social stratosphere. Ferdinand Magellan's expedition in the early sixteenth century led craftsmen, priests, barber-surgeons, laborers, artillery-men, and officers of various nationalities and social statuses into the unknown.[1] Later, in the Second Age of Exploration, the three voyages led by Captain James Cook in the eighteenth century (on the *Endeavour* 1768–71; on the *Resolution* and the *Adventure* 1772–75; on the *Resolution* and the *Discovery* 1776–79/80) or the Russian *Rurik* expedition under Otto von Kotzebue's command 1815–18, to name just a few examples, added to the crew an array of scientists of diverse European origin, artists, cartographers, and gardeners, and Pacific islander travelers and interpreters like the famous Omai and Tupia.[2] But not only these socially and culturally diverse people with their variedly held, marked, clad, and cared for human bodies could be found on the ships heading to farther shores and back. There was also an abundance of non-human bodies on board.[3] The *Endeavour*, as Anne Salmond reminds us, at one point carried for example "a motley collection of pigs, sheep, ducks, and chickens in pens on the forward deck" joining "cockroaches and other unpleasant insects, European rats, a cat, Joseph Bank's greyhound and a nondescript bitch".[4]

To avoid hunger and scurvy, European maritime expeditions brought with them various non-human animals ('animals' in the following) for the voyage to the other side of the globe:[5] the main task of Magellan's expedition had been to find an alternative route to the so-called Spice Islands of the Moluccas and their riches in cloves and other highly esteemed spices. Finally, arriving in the Moluccas, in 1531 after 27 months, the travelers were in such a devastating physical state that, as Antonio Pigafetta's account tells us, many of their trading goods – linen, scarlet cloth and crimson satin, yellow damask, Indian cloths, caps and crystal glasses, knives and scissors, combs and mirrors – had to be exchanged for not only

DOI: 10.4324/9781003331803-7

the then still exotic cocoa nuts and figs, but goats and chickens instead of pricy spices.[6] Since the voyages to the Pacific were long and salted and cured meat could only sustain a crew so far and so well, ships sailing abroad needed 'live provisions' to sustain their crew; and therefore, a substantial portion of the limited ships' space was given to living animals brought from Europe or picked up and traded for along the way. Cook's expeditions could already rely on trading posts and routes; they collected their livestock on the South African Cape. Earlier, the availability of potential live provisions was meticulously noted down in seventeenth-century navigators' accounts. When sighting unknown islands, navigational information and the geography of a new shore often swiftly gives way to a perspective on the visible living world, noting the abundance and diversity of fruits and animals. The information, explicitly collected for future travelers, thereby entailed a shopping list of edible island life, an often life-saving instrument. The English buccaneer-turned-explorer William Dampier recounts in his *A New Voyage Round the World* (1697) how close the crew of the *Cygnet*, fittingly under the command of the privateer Charles Swan, came to mutiny in March 1686: coming from the East Indies, they had "not 60 days Provisions, at a little more than half a pint of Maiz a day for each man, and no other Provision except 3 Meals of salted *Jew-fish*".[7] Since they also found "a great many Rats aboard, which we could not hinder from eating part of our Maiz" (279) and "in all this Voyage [...] did not see one Fish, not so much as a Flying-fish, nor any sort of Fowl", the situation was dire: "the men began to murmur against Captain *Swan*" (282). Drily, Dampier remarks on the sighting of the island of Guam:

> It was well for Captain *Swan* that we got sight of it before our Provision was spent [...]; for I was afterwards informed, the men had contrived to kill Captain *Swan* and eat him when the victuals were gone, and after him all of us who were accessary in promoting the undertaking of this Voyage.
>
> (283)

Mutiny and cannibalism are narrowly avoided because the *Cygnet* reaches Guam, where the Spanish colonial administration is compelled to provide them with fruit and hogs. At the next stop, Mindanao, Dampier inventories:

> many sorts of Beasts, both wild and tame; as Horses, Bulls, and Cows, Buffaloes, Goats, Wild-hogs, Deer, Monkies, Guano's, Lizards, Snakes, &c. [...] The Hogs are ugly Creatures; they have all great Knobs growing over their Eyes, and there are multitudes of them in the Woods. They are commonly very poor, yet sweet. Dear [sic] are here very plentiful in some places, where they are not disturbed. [...] The

Fowls of this Country are Ducks and Hens; Other tame Fowl I have not seen nor heard of any. The wild Fowl are Pidgeons [sic], Parrots, Parakits [sic], Turtle-dove, and abundance of small Fowls. There are Bats as big as a Kite.

(320–21)

But animals in this time are not only given culinary attention like the ugly and poor, yet sweet hogs of Mindanao described here by Dampier. Animals were also prized for their faculties and, if given, their novelty to the European eye. Ships carried working animals, for example, cats to contain vermin (such as Dampier's rats) and watchdogs, the Western epitome of a companion species.[8] Both moonlighted as pets, and, nearing the Age of Enlightenment, they were joined by increasingly 'exotic' animals from faraway islands that were not to be eaten or used for labor, but collected by naturalists and gifted to kings and queens, financiers, and patrons. Some animals even made several voyages under different captains: one of the milk-goats on Cook's *Endeavour* had, for example, already sailed successfully around the world with Captain Samuel Wallis.[9]

If we turn our gaze from the mammals to the critters, we see that the human travelers are made to feel their own body in agonizing ways by the more stubbornly 'Animal Others'. We find Adelbert von Chamisso writing about flees involuntarily taken aboard the *Rurik* in Chile and torturing the crew. In his *Reise um die Welt* (1836; published in English in 1986 as *Voyage Around the World*), Chamisso also complains about cockroaches so tenacious that they not only eat the crew's biscuit provisions but nibble on everything in reach, even on human bodies: burrowed into the ear of a sleeping seafarer, Chamisso informs us, they cause unspeakable pain.[10] While the livestock onboard potentially sustains the human body, invisible animal forces pester it in ways that compel its owners to reflect on the nature of animal agency where least suspected.[11] Men feed on animals, but animals, as every sting and nibble makes tantalizingly perceptible, also feed on men.

But even the familiar, domesticated, and controllable animals potentially represented more than an edible body and its faculties. Captain Swan, for example, retaliated for his crew's cannibalistic plotting by giving away the ship's beloved dog to the Governor of Guam, as Dampier recalls: "We had a delicate large *English* Dog: which the [Spanish] Governour did desire, and had it given him very freely by the Captain, though much against the grain of many of his Men, who had great value for that Dog" (302). As part of a victual economy, provision animals could also remind the crew of the social stratosphere that was rigidly maintained onboard, especially within the British Navy. John Duckworth, one of the captains who served as judges in the court-martial of the supposed *Bounty* mutineers who were brought back from Tahiti in 1792, apparently entertained what amounts to a 'porcine fancy'. He was known for not only farming hogs on shore but also keeping pigs on board, which he bartered with the wardroom or midshipmen's mess. When one of the pigs was swept overboard in rough weather, as a popular anecdote goes, Duckworth stammered: "Back the yards, back the yards; lower the boat, there's a pig overboard; my pig – pig – pig will be drowned." When a midshipman reminded Duckworth that the pig in question had been acquired by the crew from Duckworth a short while ago ("It is *our* pig – our poor little, new pig!"), Duckworth quickly changed mind: "What – what? *their* pig – their pig: Keep on your course [...] we must not risk – risk – risk men's live for a pig, poor thing; they can buy another!".[12]

As these short anecdotes show, onboard animals were endowed with more significance than just their nutritional value or use as working animals and pets.[13] In the following, two further assignments of animals and their bodies are of interest. These two assignments appear to have – at least retrospectively – even more value to the project of European global exploration than the survival and comfort of maritime crews, because they position animals in the realm of the political. In this political realm, animals were granted passage onboard, firstly, because they could be used as variously employed means of establishing and maintaining contact with other cultures and societies – a creaturely 'way in' – and, secondly, because they were one of the most important instruments of preparing non-European islands and continents for future European usage. What I intend to do in the following is not so much a retelling of an entangled global history of exploration, colonialism, and imperialism as animate history.[14] From a Cultural and Literary Animal studies perspective, I am interested here in the different literal and symbolic uses the traveling animal body is subjected to – or rather (as these are uses derived from textual representations) the uses that are ascribed to it and that amount to practices of inter-cultural but also inter-species contact. Because these are assignments that animals perform(ed) well precisely because the animal is more often than not reduced to 'just' a body and at the same time that body is more than just 'meat'.

5.2   Animal Encounters

When the British Captain Samuel Wallis 'discovered' Tahiti on 18 June 1767, the *Dolphin* was in urgent need of new supplies. As soon as someone could be seen on shore, the Europeans started imitating the animals they desired to eat. As George Robertson, the master of the *Dolphin*, recounts:

> We made signs to them, to bring off Hogs, Fowls, and fruit and showed them coarse cloth, Knives, Shears, Beads, ribbons etc., and made them understand that we was willing to barter with them. The method we

took to make them Understand what we wanted was this: some of the men Grunted and Cried like a Hog, then pointed to the shore – others crowed Like cocks, to make them understand that we wanted fowls. This the natives of the country understood and Grunted and Crowed the same as our people, and pointed to the shore and made signs that they would bring us off some.[15]

This scene of first contact between a British ship and Tahitian islanders stands out for two reasons: first, it is remarkable that the first thing the travelers would do, before they presented themselves as visitors or tried to make formal inquiries where they were and who they should talk to, was to ask not only for food, but specifically for pigs and poultry. Secondly, apparently, the language of animals, that is the utterances typical for these two species, functions in this scene of contact as an immediate means of communication, a language that is transmitted and – as Robertson emphasizes – *understood* by cultures formerly unknown to each other.[16] Imitating the desired animals renders the Europeans' desires comprehensible for Tahitians on shore; animals and, in this episode's *histoire*, animals soon to be only edible bodies are a means of establishing an intercultural understanding. A form of communication seems to be initiated in this encounter that carries on, albeit in a more sublimated form.

As the example of Captain Swan's gifting of the dog to the Spanish Governor of Guam shows, the presentation of an animal as a means of compliance or currying-favor was an element of European tradition and custom that was carried into the Pacific, at least where other European colonial powers were concerned.[17] This custom was translated into intercultural encounters as soon as Europeans met island societies who could not just be overrun by a Jesuit armada as happened on Guam,[18] and whose political systems appeared to be comparatively 'readable' by European seafarers. The prime example here is Tahiti. From the arrival of Captain Wallis and the hungry crew of the HMS *Dolphin* in 1767 until the major political, social, and religious changes on the island in 1827, an impressive "movement of plants and animals across cultural divides"[19] took place. As Jennifer Newell has argued, this exchange is inaugurated by Captain Wallis planting not only a British flag on the island but also a garden with peas and gifting the chiefess Purea with three guinea hens, a pair of turkeys, and a pregnant cat. It continues with the French explorer Louis Antoine de Bougainville and the famous Captain Cook, respectively, presenting the Tahitian rulers with breeding pairs of cattle, goats, sheep, and poultry.[20] In exchange, the Europeans appear to have asked so often for such a large number of pigs (and poultry, and other 'natural products') that, as Newell reminds us, when Cook sailed up the Tahitian coast on his second voyage, "the islanders could be seen running up into the hills to hide their pigs".[21]

The hogs, I suggest, are joined by at least three other species that, as anecdotal episodes show, function as central colonial means of communication[22] among humans in intercultural encounters in the Pacific: while pigs are a highly sought after but contentious foodstuff, dogs suffer the consequences of a cannibalistic discourse, and cows and goats are an instrument of imperial endeavors.[23]

As indicated above, the Christian European mariners favored pork and in most of the Polynesian Pacific Island cultures found other pork-eaters that they could barter with. Even if the European appetite weighed heavily on the islands' pig populations, when the crew of the *Dolphin* grunted and squealed, the Tahitians understood that they were looking for a feast. But the animal body as nourishment and the question of which body is to be eaten, which to be spared, which to be revered, and which to be despised, holds the potential for intercultural (dis-)agreement, possibly even conflict, as Mary Douglas' work on *Pollution and Taboo* has amply demonstrated.[24] When Magellan's expedition reached the Malay Archipelago, Pigafetta tells us, the Muslim king of the island of Ternate seemed to have been offended by the amount of hogs the Europeans were carrying onboard at the time and offered live poultry and goats if they would kill all of them:

> [H]e begged us, for love of him, to kill all the pigs which we had in our ships, for which he would give us as many goats and poultry, and to hang the dead pigs in a covered and enclosed place, so that, if perchance his people saw them, they should cover their faces in order not to see them or smell their odor.
>
> (118–19)

As Pigafetta recounts, the Europeans had previously already gifted all their 'Indian' prisoners to this king, including "three women captives, in the name of the Queen" (118), which he had asked for as a token of the Spanish king's appreciation. Now they also complied with his second demand and took the pigs below deck to slaughter them, respecting the cultural sensibility concerning the edibility and tolerability of certain animals. To ensure goodwill in a potentially fraught intercultural relationship, the Spanish–Portuguese–Italian crew of Magellan's flotilla was willing to dispose of a substantial amount of otherwise highly appreciated pork meat, not even 40 years after the Christian Reconquista of the Iberian Peninsula. The Muslim ruler on the other hand appears to have known not to ask without offering a trade: the gesture of potency also visible in his request – testing how far the Europeans were willing to go to please him – is sweetened with the offer of a substitution of the food source with living animal bodies that were esteemed edible in all cultures concerned.

Dogs, meanwhile, somehow linger on the border of discourse about the edibility of animal (and human) bodies in the Pacific. As Dampier recounts, the English dog was given to Guam's Governor by the captain of the *Cygnet* shortly *after* the crew that appreciated this dog as a companion animal had discussed the possible slaughter and eating of this captain. This remark hints at the way that dogs function as figures of substitution themselves. In her eponymous study, Salmond tells of the trial of the 'cannibal dog' on the *Discovery* during Cook's third voyage. The dog in question was from New Zealand, caught by an officer of the *Discovery* who intended to bring it back to Britain as a present for a benefactress.[25] In his absence, his shipmates one day decide to put the 'savage' dog on trial for being of Polynesian decent and, so this logic goes, therefore a cannibal. On Cook's second voyage, as George Forster retells it in his *A Voyage Round the World*, the shocked Europeans had witnessed cannibalism perpetrated on some of the crew by Maori of the Northern Island's shore (1: 511–18).[26] Since this dog had already bitten the crew several times, the imagined link between Polynesia and cannibalism was enforced, the dog was declared guilty and executed. However, the course of action then sways first towards the cannibalistic itself and later towards the carnivalesque: the body of the dead dog was dressed to be eaten as a meal. A portion was set aside for the owner of the dog, who, upon boarding the ship again, was also decorated with the dog's hide in an apparent mockery of Maori royal dress that was fashioned out of dog pelts. The symbolic and the victual meet in this meaty anecdote: after some time of angry deliberation with himself, the former European owner of the dead Polynesian 'cannibalistic' dog joins his comrades and, possibly still with the dog hide on head and shoulders in costume of either a Maori ruler or a dog, joins the canine meal. It is telling that among all the dogs that Cook's expeditions carried on board (Forster tells of at least 30 dogs onboard the *Resolution* and the *Adventure* when they sailed, 1: 387), the consumed dog is brought on board as a foreign – and, since it is intended as a gift for a benefactress back in England, clearly exotic – body. The whole episode can be interpreted as a symbolic act whereby a Maori act of cannibalism perpetrated on some of Cook's men from the second voyage is demonstratively avenged on this third voyage by sentencing a dog stand-in for 'Polynesians'.[27] This would explain the trial held and the official sentence to be executed, as Salmond argues.[28] However, the episode can also be regarded as an act of social and cultural transgression, as dogs were usually not eaten in England at this time,[29] and dog meat in the Polynesian cultures encountered at the time was a prerogative of the ruling classes. Thus, by eating the flesh of a dog, the dog-eaters of the *Discovery* constructed an intersectional transgression against (at least) two 'taboos'. Salmond interprets this, not unproblematically, as a 'going Polynesian' of the British sailors.[30] Further interpretations see this anecdote, with its strange charade of the dog-as-human/human-as-dog, as containing a treatment of the Europeans' own cannibalistic desire.[31] Then again, there are ample examples of dogs being eaten aboard ships, especially on the Cook voyages, and they are often eaten as a special treat or remedy, as when the elder Forster sacrifices one of his favorite dogs for a broth, that is, to reconstitute the health of the then seriously ill Captain Cook (Forster 2: 3, 1: 234).[32]

The dog moreover functions as a facilitator of discourses *on* cannibalism. First and foremost, this can be illustrated by a passage in Forster's *Voyage*. Forster first points to the time "spend on the education of dogs, that they acquire those eminent qualities which attach them so much to us" (1: 235) as being responsible for the European "aversion to dog-flesh".[33] He then differentiates between these educated, endearing European dogs and the Polynesian (or rather, in this case, Maori) ones, which are "the most stupid, dull animals imaginable" (1: 235), almost like sheep. By likening them to sheep, Forster moves the Polynesian dogs closer to being a victual. Actually, the whole passage starts with an anecdote about a ship's dog being eaten and tasting "so exactly like mutton, that it was absolutely undistinguishable" (1: 235).[34] If the meat tastes like mutton and the dogs of the Pacific are dull as sheep, why *not* eat them, Forster seems to argue. But then Forster moves to his cannibal argument, a theorizing about instinct and familiarization, that would later in his *Voyage* lead him to an argument about the place of cannibalism in a universal cultural history of mankind (1: 511–518):[35] once familiarized with eating the flesh of other dogs, the (Polynesian) dogs will produce offspring already instinctively endowed with a cannibalistic appetite for dog meat. The dogs from New Zealand, being fed "on the remains of their masters' meals" eat dog bones and, as Forster's experiment with "a young New Zeeland puppy on board" shows, "eagerly" devour dog meat, "while several others of the European breed taken on board at the Cape, turned from it without touching it" (1: 236). A few days later, the same puppy earns the term 'cannibalistic dog' by not only eating a stillborn Dachshund, but also by licking a seaman's bloody finger and trying to bite it off (1: 243). Cannibalism viewed through the lens of 'dog discourse' as purported by Forster thus connects the eating of canine and human bodies either by dogs or by humans, and it is not always clear, which is which. Forster thereby not only relativizes and historicizes anthropophagy among the Maori via his substitution discourse around the cannibal dog but, by discussing equally their edibility and eating-habits, blurs the anthropocentrically established lines between animal and human bodies, at least the one that is drawn on the terrain of edibility.

While pigs and dogs were species widely known in the Pacific before the arrival of Europeans – as can be seen by the deadly uneasiness that

Magellan's pigs encountered in the king of Ternate, by the Tahitians understanding the human hog noises, and the Polynesian dogs picked up all around the Pacific islands by European voyagers – bovines and goats, not as easy to transport in the earlier inter-island expansion, were a novelty in the Pacific east of the Malay influence. Goats were first introduced by the Spanish. Strategizing the long voyage across the Pacific, they replenished mostly uninhabited Pacific islands with goats as a meat supply for passing ships on their way to Manila from Lima and other west coast ports of the Americas. Later exploration leaders like Bougainville, Cook, and George Vancouver seemed to have made a point of carrying goats and cows alike with them. On his third voyage, Cook's agenda explicitly entailed stocking the Pacific islands for posterity.[36] When he was made aware that a Spanish ship had called at Tahiti shortly after his last visit there, and that the Spanish had made an impression on the islanders among other things by bringing cattle to the island – before Cook was able to deposit his bovine presentation – he (and his crew) were devastated: James King, officer and astronomer on the *Resolution*, writes of "disappointment & vexation".[37] On seeing the Spanish bull at Matavai, Cook concedes that he was "a finer beast [...] than I hardly ever saw";[38] but it was a lone bull. After presenting the animals he brought with him that day – a peacock, a hen, a turkey, a drake, and ducks – he therefore sends "three Cows I had on board to the Bull"[39] the next day. He also puts ashore another bull, a horse, a mare, and sheep, and then concludes:

> I now found my self lightened of a very heavy burden, the trouble and vexation that attended the bringing of these Animals thus far is hardly to be conceived. But the satisfaction I felt in having been so fortunate as to fulfill His Majestys design in sending such usefull Animals to two worthy Nations sufficiently recompenced me for the many anxious hours I had on their account.[40]

The satisfaction Cook mentions might also derive from knowing that the British cows would surely in time dilute any livestock breeding line that the rival colonial power had wished to install on Tahiti.

As Inga Clendinnen has argued for the Aztecs vis-à-vis Spanish horses, the new animals, although at first sight potentially frighteningly exotic for non-European peoples, could easily be identified as *animals* by their bodily gestalt.[41] And if the European behavior was any indication, all of these exotics could be eaten. But, as the islanders quickly realized, they also carried meaning for the European visitors: first of all, as representative gifts of these new foreign friends and their faraway rulers, the animals held comparative value. Cook's crew was rather disappointed when they realized that their presentation of cows fell flat because the act of ceremoniously acquainting the Tahitians with cattle had fallen to the Spanish. In the eyes of the Tahitians, the British (or rather South African) bovines did not seem to compare very well with the Spanish (or South American) cattle.[42] As most of the Polynesian islanders also brutally learned, misunderstandings about the meaning and value the Europeans adhered to these animals and their bodies could have grave consequences for the islands and their human as well as animal populations.

A well-known episode of British–Polynesian relations concerns the theft of a goat from Cook's ship at Moorea or Eimeo, as Cook called that part of the Society Islands, and Cook's retaliation against the supposed thieves. As Moorea's chief Mahine had asked Cook earlier for goats, which Cook, having other islands to be stocked in mind, had denied him, Mahine became the prime suspect. Cook offered to send to a neighboring island for other goats, but meanwhile another goat, this one pregnant, was stolen. The first one was then returned with the explanation that it had only been taken as retribution for plants and fruit stolen by the Europeans before. While Cook accepted this explanation, he then sent 40 men into the depth of the island for the retrieval of the second goat. They destroyed canoes, burnt houses, and plundered their inhabitants – and while the goat was not found, the islanders tried to appease the Europeans with plantain trees. However, the next day Cook, incensed by ill advise from his Polynesian advisers (Omai and two older men), sent off a rampage party destroying almost every canoe on the island, setting Moorea in many ways back for years to come.[43] Cook not only employed a radical tool of colonial power – the punitive expedition – in a per-se pre-colonial setting; an episode which was afterwards much regretted and criticized, particularly by Cook's own men. His lesson in property rights and 'British superiority' to the Society Islanders also taught them the price of a goat to be almost the entirety of an island's vital means of transportation, economy, and warfare.[44] The goat was eventually returned, alive, but its price for the British was a horrendous reputation among the Society Islanders in a situation of Inter-European rivalry staged in the Pacific.

5.3   Imperial Animal Terra-Forming

In the eighteenth century, the European nations were already competing for influence in the Pacific. Facilitating the first encounter of Pacific Island peoples with exotic animals of more or less European origin was one of the strategic steps taken. But the animals populating the ships' decks and storage rooms served more than this symbolic entrée into (pre-)colonial one-upmanship. A greater plan was at work: the animals were pioneers in a project of adjusting Pacific landmasses for future European use. More than just the replenishment of passing ships with living meat-packages, the stocking of islands with European animals aimed at forming what Alfred Crosby has called "Neo-Europes".[45] The ecological imperialism elucidated by the Europeans' proclivity for migrating overseas, so Crosby's leading argument, is mainly facilitated by domesticated animals.[46] By the time Europeans ventured into the Pacific, ecological imperialism was fully fledged – it was no longer a biogeographical byproduct of early human colonists carrying their domestic livestock with them wherever they went, setting them free so they had their hands free to build homes and societies (and kill the indigenous population) and then losing control over the spread of animal populations entirely foreign to the ecosystems they were transported to. "We had not only endeavoured to leave useful European roots in this country", Forster writes when recounting a visit with the captains Cook and Furneaux in 1773 to East Bay and Grass Cove on New Zealand's North Island, "but we were likewise attentive to stock its wilds with animals, which in time might become beneficial to the natives, and to future generations of navigators" (1: 221). Furneaux had already set free a boar and two sows at another spot, Cannibal Cove; the landing party now marooned "with the same view, [...] a pair of goats, male and female, which we left in an unfrequented part of East Bay" (1: 221). The animals were labeled "our new colonists" and as such their well-being was of much importance.[47] They were intentionally set free in areas where, so Cook and his men hoped, they would "remain unmolested by the natives" (1: 221). These 'natives' were considered to have too much of an "inconsiderate and barbarous temper" (1: 221) to have "any reflection on the advantages which future ages might reap from the propagation of such a valuable race of animals" (1: 221–22).

Earlier, Forster and Cook had already left "five tame geese" originally from the Cape of Good Hope in a bay they then named 'Goose Cove'. The cove was "the most convenient place for that purpose since there were not inhabitants to disturb them" (1: 176). On the South Island Cook, Dr Sparrman, the older and the younger Forster dispatched for another such clandestine mission and set free "two sows and a boar, with three cocks and two hens" (1: 507). Here Forster writes again that the spot was picked because it was unlikely to be frequented by New Zealanders and "the animals would be left to multiply their species without any molestation" (1: 507). Some islanders in a canoe had seen them entering the bay, but as Foster assures his readers, they "probably would not suspect that we were come on so particular an errand" (1: 507). This errand's aim is: "If therefore the southern isle of New Zeeland should in course of time be stocked with hogs and fowls, we have great reason to hope that the care with which we concealed them in the woods, has been the only means of preserving the race" (1: 507).

Unbeknownst to Forster, Cook, and others at the time, the non-endemic species they set free in quiet island coves held the potential to overthrow – in time – whole island ecosystems, more often than not in an entirely unpredicted way.[48] In our age of shrinking biodiversity and trophic cascades, it appears menacing how the European men sneaked into secluded island bays and released biogeographical time bombs with cheery words:

> We set them on shore, [...] pronouncing over them the *crescite & multipliciamini*, for the benefit of future generations of navigators and New Zeelanders. There can be little doubt indeed, but that they will succeed in these secluded spots, and in time spread over the whole country, answerable to our original intention.
>
> (1: 176, emphasis in text)

But as Forster emphasizes in his Preface to the *Voyage*, anticipating the third Cook expedition, the "present of new domestic animals" to the islands of the Pacific was regarded as a benevolent project conducive not only to the future exploration and presence of Europeans in the Pacific but also to the island populations themselves – and it was supposed to further the perfectibility of mankind:

> The introduction of black cattle and sheep on that fertile island, will doubtless increase the happiness of its inhabitants; and this gift may hereafter be conducive, by many intermediate causes, to the improvement of their intellectual faculties. And here I cannot but observe, that considering the small expence at which voyages of discovery are carried on, the nation which favours these enterprizes is amply repaid by the benefit derived to our fellow-creatures. I cannot help thinking that our late voyage would reflect immortal honour on our employers, if it had no other merit than stocking Taheitee with goats, the Friendly Isles and New Hebrides with dogs, and New Zeeland and New Caledonia with hogs.
>
> (1: xvii–xviii)

Today, many of these traveling animals of the First and Second Age of Discovery and Exploration have left 'their mark', so to say, on the islands of the Pacific. It is not just the ubiquitously invasive rat that has conquered the world by ship. New Zealand's feral pig population, responsible for damaged pastures and crops, killed lambs, the destruction of endemic species' habitats, and thereby the demise of the kakapo, for example, can be traced back to the porcine colonists the Cook expeditions placed in New Zealand's bays. They are therefore often called 'Captain Cookers'.[49] Four Spanish goats marooned by Juan Fernandez in 1540 on the island named after him proliferated so well that they not only sired a new subspecies of goats but, in collaboration with rats, mice, and other cattle, nearly ruined the island's existing ecology – and found their fictionalization in Daniel Defoe's *Robinson Crusoe*.[50] The four bulls and eight cows gifted by Vancouver to the Hawai'ian king Kamehameha I spawned Hawai'ian feral cattle once numbering in the ten thousands, endangered native flora, provoked land erosion, and instigated a Hawai'ian cowboy culture.[51] On Guam, the game that hunting-affine Spanish gentlemen brought to the island has given way to one of the worst trophic cascades known to biogeography: in the 1990s, the brown tree snake, introduced mid-twentieth century, had diminished the island's bird fauna to almost nothing, and next came small mammals and bats – the only profiteers appear to be millions of spiders.[52]

5.4   Conclusion

As this analysis of anecdotal accounts of animals in the Pacific has shown, the pigs, goats, cows, dogs, geese, etc., onboard European ships setting sail to the other side of the world were not 'only' surplus bodies on board, live provisions, or furry helpers. Animals were traveling bodies. Their value to the maritime enterprises derived from the fact that the animal was conceived of as a body of diverse, but possibly interculturally shared or at least understood significance. It was therefore a body whose traveling facilitated practices of discovery, of cultural encounter, of imperial and colonial politics, and of a European enlightenment discourse – at least where discourses on cannibalism and the philosophy of history became entangled. In fact, as mediums for the communication and negotiation of social norms and values as well as symbols of power structures and tools of facilitating planned colonial and imperial futures, these 'on board animals' themselves can be understood as agents of exploration, of intercultural encounter, and of the eventual imperial colonization and ecological change wrought on the islands of the Great Ocean.

Notes

1 Nancy Smiler Levinson, *Magellan and the First Voyage Around the World* (New York: Clarion Books, 2001), p. 39–50.
2 See, for example, Richard Connaughton, *Omai. The Prince Who Never Was* (London: Timewell Press, 2005); and Joan Druett, *Tupaia. Captain Cook's Polynesian Navigator* (Santa Barbara: Praeger, 2011).
3 Non-human body parts were of course as commonly used in seafaring as in everyday life. But leather, animal hair, and similar animal products do not interest me in the following; it is the living non-human animal on board, defined as belonging to a species by the definition of its outward bodily form, that is of interest to the study of traveling bodies. For the long history of morphological concepts of 'species' see Georg Toepfer, "Art", in Georg Toepfer, *Historisches Wörterbuch der Biologie. Geschichte und Theorie der biologischen Grundbegriffe*, vol. 1 (Stuttgart: J. B. Metzler, 2011), pp. 61–131.
4 Anne Salmond, *Two Worlds. First Meetings Between Maori and Europeans 1642–1772* (Auckland: Viking, 1991), p. 103.

5 Seafaring peoples all over the globe have had animal companions and live provisions accompanying them to new shores; above all the Polynesian exploration of the Pacific (conservatively held to have started around 1500 BC from the Malayan Archipelago, New Guinea and/or Taiwan) contained at least dogs, chickens, and pigs as fellow travelers. The pig or dog dispersal and genomes in the Pacific can actually be used to study the history of human migration, see, for example, Keith Dobney et al., "The Pigs of Island Southeast Asia and the Pacific: New Evidence for Taxonomic Status and Human-Mediated Dispersal", *Asian Perspectives*, 47, no. 1 (2008): 59–74; and K. Greig et al., "Complex History of Dog (Canis familiaris) Origins and Translocations in the Pacific Revealed by Ancient Mitogenomes", *Scientific Reports*, 8, no. 1 (2018): 9130.

6 Antonio Pigafetta, *Magellan's Voyage. A Narrative Account of the First Circumnavigation*, trans. R. A. Skelton (New York: Dover Publications, 1969), pp. 115–18. Further references to this edition are included parenthetically in the text.

7 William Dampier, *A New Voyage Round the World* (London: James Knapton, 1697), p. 279, emphasis in text. Further references to this edition are included parenthetically in the text. Dampier is here probably speaking of a species of the *Epinephelus* genus since re-named Goliath grouper. It is unclear why the fish was previously called 'jew-fish'. For theories see Avishay Artsy, "How the Jewfish Got Its Name" (2015), *Jewniverse, From the Jewish Telegraphic Agency*, www.jta.org/jewniverse/2015/how-the-jewfish-got-its-name [19 May 2022].

8 For the relevance of dogs as "a species in obligatory, constitutive, historical, protean relationship with human beings" see Donna Haraway, *The Companion Species Manifesto. Dogs, People, and Significant Otherness* (Chicago: Prickly Paradigm Press 2003), pp. 11–12.

9 Salmond, *Two Worlds*, p. 103.

10 Adelbert von Chamisso, *Reise um die Welt* [1836] (Berlin: Die Andere Bibliothek, 2012), p. 110.

11 See Chamisso, *Reise um die Welt*, pp. 234–36: "Es hat etwas Unheimliches, etwas Wundergleiches, wenn die Natur einer solchen untergeordneten Art, deren Individuum als ein unmächtiges Nichts erscheint, durch die überwuchernde Anzahl derselben [...] zu einer unerwarteten Übermacht verhilft. Dem Menschen verborgen, entziehen sich seiner Einwirkung die Umstände, welche die Vermehrung und Abnahme jener Geschlechter bedingen; sie erscheinen und verschwinden. Dem Spiele der Natur sieht er unmächtig staunend zu." ["There is something weird, something close to the miraculous about it when nature assists such a subordinate species, any individual member of which appears as a powerless nothing, through their swelling numbers [...] to attain an unexpected superiority. Concealed from man, the conditions that affect the increase and decrease of those species are beyond his control; they appear and disappear. He watches the play of nature powerless and astonished."] (Adelbert von Chamisso, *A Voyage Around the World with the Romanzov Exploring Expedition in the Years 1815–1818 in the Brig Rurik, Captain Otto von Kotzebue*, trans. and ed. Henry Kratz (Honolulu: University of Hawai'i Press, 1986). For a concrete suggestion of animal agency see Helen Steward, "Animal Agency", *Inquiry*, 52, no. 3 (2009): 217–31; for the history of the concept see Chris Pearson, "History and Animal Agency", in Linda Kalof, ed., *The Oxford Handbook of Animal Studies* (Oxford: Oxford University Press, 2014), pp. 240–57.

12 Mrs. Cornwell Barron-Wilson, *Memoirs of Miss Mellon, Afterwards Duchess of St. Albans* (London: Remington & Co, 1886 [New Edition]), pp. 300–01, emphasis in the text.

13 For a history of animal-assisted activities and working animals see Margo DeMello, *Animals and Society. An Introduction to Human-Animal Studies* (New York: Columbia University Press, 2012), pp. 194–200. For more on maritime pets see Sari Mäenpää, "Sailors and Their Pets: Men and Their Companion Animals Aboard Early Twentieth-Century Finnish Sailing Ships", *International Journal of Maritime History*, 28, no. 3 (2016): 480–95; and Vittoria Traverso, "The Little-Known History of Seafaring Pets" (2018), *Atlas Obscura*, www.atlasobscura.com/articles/little-known-history-seafaring-pets-dogs-cats-chickens-war-exploration [19 May 2022].

14 Pascal Eitler and Maren Möhring, "Eine Tiergeschichte der Moderne. Theoretische Perspektiven", *Traverse: Zeitschrift für Geschichte*, 3 (2008): 92–105; and Gesine Krüger, "Tiere und Imperium – Animate History postkolonial: Rinder, Pferde und ein kannibalischer Hund", in Gesine Krüger, Aline Steinbrecher, and Clemens Wischermann, eds., *Tiere und Geschichte. Konturen einer Animate History* (Stuttgart: Franz Steiner, 2014), pp. 127–53.

15 Oliver Warner, ed., *An Account of the Discovery of Tahiti: From the Journal of George Robertson Masters of HMS Dolphin* (London: Folio Press, J. M. Dent, 1973), p. 21.

16 Jennifer Newell mentions, though, that the Tahitians had at least heard of previous contacts of other islanders with European travelers (*Trading Nature. Tahitians, Europeans & Ecological Exchange* (Honolulu: University of Hawai'i Press 2010), pp. 28–29). Connaughton depicts the later 'official' first encounter on the shore being mainly dominated by incomprehensible speech acts on both sides that are interpreted as benevolent simply because of the lack of violence involved (*Omai*, p. 8).

17 Felicity Heal, *The Power of Gifts. Gift-Exchange in Early Modern England* (Oxford: Oxford University Press, 2014), pp. 155–200.

18 Robert F. Rogers, *Destiny's Landfall. A History of Guam* (Honolulu: University of Hawai'i Press 1995), pp. 41–57.

19 Newell, *Trading Nature*, p. ix.

20 Newell, *Trading Nature*, p. 7; see also Krüger, "Tiere und Imperium", pp. 145–52.

21 Newell, *Trading Nature,* p. 13. George Foster elaborates on the *Resolution*'s stay at Tahiti in August 1773: "Notwithstanding the friendly reception which we met with on all sides, the natives were very anxious to keep their hogs out of sight, and whenever we enquired for them seemed uneasy, and either told us they had none, or assured us they belonged to Aheatua their king." (*A Voyage Round the World, in His Britannic Majesty's Sloop, Resolution, commanded by Capt. James Cook, during the Years 1772, 3, 4, and 5*, vol. 1, vol. 2 (London: B. White, J. Robson, P. Elmsly, G. Robinson, 1777), p. 287). Further references to this edition are included parenthetically in the text; the first number indicating the respective volume.

22 Communication is here understood in its broad sense as an act of conveying meanings from one entity or group to another through the use of mutually understood signs, symbols, and semiotic rules. Communication is as necessary to enforcing barter as it is to implementing colonial power systems. Based on the material here discussed, it can be argued that animals function as mutually understood signs, symbols, and even convey semiotic rules.

23 Krüger pays special attention to horses, but since their value as a means of transportation and a means of conquest has been given ample academic atten-

tion and as Krüger has already described their importance for pre- and colonial endeavors, I will refrain from discussing the role of horses here. See Krüger, "Tiere und Imperium", pp. 146–47, and in reaction to Tsvetan Todorov: Inga Clendinnen, "'Fierce and Unnatural Cruelty': Cortés and the Conquest of Mexico", *Representations*, 33 (1991): 65–100 (at pp. 82–83); and Gananath Obeyesekere, *The Apotheosis of Captain Cook. European Mythmaking in the Pacific* (Princeton: Princeton University Press, 1997), pp. 18–19.

24 Mary Douglas, *Purity and Danger: An Analysis of Concepts of Pollution and Taboo* (London: Routledge and Keegan Paul, 1966).

25 Anne Salmond, *The Trial of the Cannibal Dog. The Remarkable Story of Captain Cook's Encounters in the South Seas* (New Haven: Yale University Press, 2003), p. 1.

26 For a discussion of this incident see Gananath Obeyesekere, *Cannibal Talk. The Man-eating Myth and Human Sacrifice in the South Seas* (Berkeley: University of California Press, 2005), pp. 30–36.

27 For this incident see Salmond, *The Trial of the Cannibal Dog*, pp. 2–3; and Jürgen Goldstein, *Georg Forster. Voyager, Naturalist, Revolutionary*, trans. Anne Janusch (Chicago: The University of Chicago Press, 2019), pp. 66–67.

28 Salmond, *The Trial of the Cannibal Dog*, pp. 4–5. For the European history of prosecutions of offending animals see Edward P. Evans, *The Criminal Prosecution and Capital Punishment of Animals* (London: William Heinemann, 1906).

29 Krüger, "Tiere und Imperium", p. 150.

30 Salmond, *The Trial of the Cannibal Dog*, pp. 8–9.

31 Krüger, "Tiere und Imperium", p. 150.

32 See also Krüger, "Tiere und Imperium", p. 151.

33 Forster roots this aversion in religious taboo, too, calling it first a "Jewish aversion to dog-flesh" (1: 235).

34 Forster explains the canine meal by elaborating that the officers had lost their taste for provisions of salted fish due to the hearty New Zealand diet they had enjoyed on shore (1: 234).

35 See also the later essay "Über Leckereyen" [On Delicacies] (Georg Forster, "Über Leckereyen", in Deutsche Akademie der Wissenschaften zu Berlin, ed., *Werke. Sämtliche Schriften, Tagebücher, Briefe*, vol. 8, pp. 164–18); and Goldstein, *Georg Forster*, pp. 63–66, 68.

36 Nicholas Thomas, *Cook. The Extraordinary Voyages of Captain James Cook* (New York: Walker & Company, 2003) p. 347.

37 Thomas, *Cook*, p. 334.

38 Thomas, *Cook*, p. 336.

39 Thomas, *Cook*, p. 336.

40 Thomas, *Cook*, p. 336.

41 Clendinnen, "Cortés", p. 82.

42 Thomas, *Cook*, p. 334.

43 Thomas, *Cook*, p. 345; see also Connaughton, *Omai*, pp. 241–43.

44 It also entailed the potential for drastic corporeal punishment: when on the way back from Moorea a stowaway and thief was discovered onboard the *Resolution*, Cook's radical stance lingered on in his orders to alter the body of the man by shaving his head and removing his ears. As Connaughton remarks, one of Cook's lieutenants intervened and reduced the mutilation to "the token removal of an ear lobe" (*Omai*, p. 243).

45 Alfred Crosby, *Ecological Imperialism. The Biological Expansion of Europe, 900–1900* (Cambridge: Cambridge University Press, 1986), p. 2. See for the

following also Mira Shah, "Aotearoa. Tierpolitiken in Neuseeland", in Roland Borgards, Lena Kugler, and Mira Shah, *Pazifische Passagen. Ein Insularium des Großen Ozeans* (Göttingen: Wallstein, 2022).

46 Crosby, *Ecological Imperialism*, p. 173.

47 Forster details the troubles of keeping animals alive and well en route to the Pacific and deliberates on sheep suffering terribly from scurvy (1: 145).

48 The Global Invasive Species Database attests to the immense impact imported species had and still have on the endemic flora and fauna, especially on islands in the Pacific. As biogeographical research has shown, their introduction often results in trophic cascades, which concerns not only competing indigenous species but which affects the whole ecosystem by the interplay of various natural agents. See John Terborgh, "The Trophic Cascade on Islands", in Jonathan B. Losos and Robert E. Ricklefs, eds., *The Theory of Island Biogeography Revisited* (Princeton: Princeton University Press, 2010), pp. 116–38; and Invasive Species Specialist Group (ISSG), *Global Invasive Species Database*, www.iucngisd.org/gisd/ [30 Sept. 2019].

49 C.M.H. Clarke and R.M. Dzieciolowski, "Feral Pigs in the Northern South Island, New Zealand: I. Origin, Distribution, and Density", *Journal of the Royal Society of New Zealand*, 21, no. 3 (1991): 237–47; and Allan Gillingham, "Pigs and the Pork Industry", in *Te Ara – the Encyclopedia of New Zealand*, www.TeAra.govt.nz/en/pigs-and-the-pork-industry/print [20 May 2022].

50 Roland Borgards, "Die Legende vom Kampf der Ziegen mit den Hunden auf der Isla Juan Fernández", in Ute Holl, Claus Pias, and Burkhardt Wolf, eds., *Gespenster des Wissens. Für Joseph Vogl* (Zürich: diaphanes, 2017), pp. 55–60; Roland Borgards, "Ziegen, Hunde und Kaninchen. Die Geschichten der Isla Robinson Crusoe, ehemals Isla Juan Fernández, anfangs Isla Más a Tierra", in Roland Borgards, Lena Kugler, and Mira Shah, *Pazifische Passagen. Ein Insularium des Großen Ozeans* (Göttingen: Wallstein, 2022).

51 John Ryan Fischer, "Cattle in Hawai'i: Biological and Cultural Exchange", *Pacific Historical Review*, 76, no. 3 (2007), pp. 347–72.

52 Mark Jaffe, *And No Birds Sing. The Story of an Ecological Disaster in a Tropical Paradise* (New York: Simon & Schuster, 1994); Mira Shah, "Guam. Von Menschen, Vögeln und Krankheiten", in Roland Borgards, Lena Kugler, and Mira Shah, *Pazifische Passagen. Ein Insularium des Großen Ozeans* (Göttingen: Wallstein, 2022).

6 Mary Wollstonecraft and the Body of Her Letters, or

The Traveler Lost and Found in Scandinavia

Michael Meyer

6.1 Introduction

The present contribution goes beyond criticism of Wollstonecraft's *Letters Written during a Short Residence in Sweden, Norway, and Denmark* (1796)[1] in its coverage of more primary material and its analysis through phenomenology. Wollstonecraft's travelogue follows the sublime and the picturesque as well as the enlightened and sentimental traditions, the latter of which foregrounds the body and senses.[2] Among the critics of the travelogue, John Whale and Elizabeth Bohls have the most pronounced focus on the body. Bohls stresses Wollstonecraft's innovative revaluation of the body as the foundation of aesthetics and politics against Edmund Burke's conservative dismissal of the body and a disinterested landscape aesthetic based on visual and social distance.[3] She notices that Wollstonecraft presents herself as a gendered and embodied observer located in and looking at peopled landscapes rather than a detached observer enjoying an empty landscape.[4] Wollstonecraft is often concerned with the impact of the harsh environment on the lives of the poor, which shapes their use and perception of the land, juxtaposing the elitist picturesque with lower-class experience in an "anti-aesthetic turn".[5] Bohls mainly stresses Wollstonecraft's new, gendered aesthetic as an integrative movement in a historical context, whereas Whale foregrounds the divisive aspects of the embodied self because the sympathy that binds the self to both nature and society is also conducive to disaffection and alienation.[6] I will focus on the more specific processes of embodied traveling, include the ominous beginning and ending of Wollstonecraft's journey omitted from the travelogue, and argue that the embodied self both suffers from and is responsible for her alienation from herself, English, and Scandinavian societies.

Critics dealing with the travelogue, published in 1796, have neglected the beginning of the journey reported in her private letters to her detached lover Gilbert Imlay, selected and posthumously published by her husband William Godwin in 1798.[7] The private letters offer a frame to the travelogue of the "suffering traveler"[8] because they present the delayed outset of her journey, which confronts her with as many difficulties as her return

DOI: 10.4324/9781003331803-8

to England. To my knowledge, no critic applied phenomenology to the travelogue as an approach that focuses on embodied experience, and more particularly, Bernhard Waldenfels's *Phenomenology of the Alien*,[9] which scrutinizes the tension between the *experience* of self and Other and its *representation* within or beyond the self's cultural framework.

Why does Waldenfels's phenomenology form a legitimate and useful approach to this historical travelogue and others which pay attention to embodied experience? Eighteenth-century sentiment and contemporary phenomenology nicely complement each other as they both address the relevance of embodied *sensation* as a basis of understanding. Concerning travel writing, the postmodern phenomenology of encountering self and other goes far beyond the theory and practice of sentiment, which informs Wollstonecraft's travelogue to some extent. In a nutshell, the eighteenth-century ideals of sensibility consider sensation and sympathy as the basis of a deep and refined relationship of the individual to nature, to other human beings, and to him- or herself. At the same time, enlightened critics have warned of the excess of sensibility and demanded restraint through delicacy, self-control, 'masculine' reason, and middle-class morals.[10] Waldenfels neither focuses on class and gender nor the display and excess of sentiment, but rather on the basic process of the embodied experience and understanding of self and other. However, Waldenfels would consider categories such as class and gender as sociocultural factors that inflect understanding. He scrutinizes the enabling and limiting factors of the self's experience of and response to the appeal of the 'alien' Other. In these respects, the *Phenomenology of the Alien* yields results that go beyond the historical contextualization of Wollstonecraft's travelogue by many critics. The following analysis situates the phenomenological analysis of the travelogue in its historical context.

Wollstonecraft as an embodied traveler literally and metaphorically struggles to find her position and her way. Her alienation from herself and the Other begins close to home and becomes more severe on the threshold between home and abroad. She tries to contain her alienation abroad through resisting the Other's challenge of her preconceptions. She assimilates the Other, reaffirming sociocultural boundaries within and not beyond her conceptual framework of Enlightenment. She takes care to position herself outside and above foreign society and tries to establish a sentimental bond with nature. In the following, an outline of a few core concepts from phenomenology precedes the analysis of selected passages about finding – and losing – her position as a writer and traveler.

6.2 Approach: Phenomenology of the Alien

Phenomenology addresses the ambiguous connection between embodied experience and understanding as well as self and Other in a nuanced way.

Traveling in foreign countries and encountering foreigners pose the difficulty to come to terms with one's potentially alienating experience. For
Waldenfels, the 'extraordinary alien' (the Other with a capital 'O') is a
liminal phenomenon that comes into existence through establishing a self
and an order with boundaries that relegate the alien to 'the other side'.
The alien is more than a relative other that can be fully recognized and
appropriated because the alien eludes our grasp.[11] However, "alienness
begins in one's own house"[12] as one's lived body evades one's will and
consciousness, and one's belonging to a culture is never complete and fully
accessible.[13] In the encounter with the Other, Waldenfels distinguishes
pathos from response. Pathos is something that happens to us: the affection by the Other designates a pre-reflective, embodied experience of a
sensation, of being seen or touched, affected as a disturbing and therefore
alienating phenomenon. This phenomenon appeals to our attention before
it is understood as being something or someone.[14] The affected self "is
literally subject to certain experiences" like a "patient" before he or she
becomes a "respondent".[15] Response means the reaction to being affected
in the shape of attention and answer to the call of the Other. The Other
presents "an appeal that is directed *at someone* and a claim or pretension
to something".[16] This "situationally embodied call"[17] one needs to behave
to as a singular event before one recognizes the question and develops an
answer. For example, if a stranger approaches us, we need to relate to
the situation before we may notice what he or she asks us for, and in our
response of giving information, we take it for a fact that the appeal for
something is a question for help in terms of directions. Our intentional
answer neither contains the initial ambiguity of the situation, the unease
of being addressed by someone unknown, or simply the disruption of our
train of thoughts nor what the stranger actually felt, thought, or wanted
over and above directions, if these were on his or her mind. One effaces
the *"responsive difference"*[18] between what one *responds* to and how one
answers if one reduces the alien to a meaningful difference according to a
familiar norm rather than being aware of the limitations of one's answer.
An appropriate response to the alien requires a creative rather than a repetitive answer.[19] The gap between pathos and response can be revealed in the
fissure between the answer to an Other and the process of trying to make
sense, which always lags behind the potentially disruptive event.[20]

The *experience of a stranger* can turn into *the estrangement of experience*, as the alien "might alienate us from ourselves. Hence the per-
petual motivation to resist, avoid, or assimilate the alien".[21] Waldenfels's
sophisticated differentiation of the process of experiencing and understanding encounters with alien individuals and cultures allows us to
notice and analyze the complexity and complications of Wollstonecraft's
Scandinavian journey.

6.3 Analysis: Finding a Voice to Represent and Communicate Experience

Wollstonecraft's "Advertisement" or preface that precedes the travelogue registers a tension between her memory of the journey and the ordering of her thoughts for publication, her experience, answer, and communication: "I, therefore, determined to let my *remarks and reflections* flow unrestrained, as I perceived that I could not give a *just description* of what I saw, but by relating the *effect* different objects had produced on my *mind and feelings*, whilst the *impression* was still *fresh*" (*A Short Residence* 241, my emphasis). Rather than writing "stiff and affected" letters, the traveler considers personal "desultory letters" that appeal to the reader's "attention by acquiring our affection" (241). The term "desultory" suggests *discursive* wandering as a correlative of her external journey and her wandering mind.[22] Embodied sensibility forms the basis of both her experience *and* its communication as a sequence of affect and response on the part of the writer as well as the reader. In her preface, Wollstonecraft devises a phenomenology *avant la lettre*, promising to render her holistic experience as closely as possible rather than imposing a reductive answer on the alien and on her readers.

6.3.1 *On Not Setting Out: Frustrated Motion and Emotion*

It is hard to imagine for the modern traveler, let alone the tourist, the hardships earlier travelers had to face. Mobility, shelter, and sustenance could not be taken for granted. Wollstonecraft was often struggling to find her way, to get a guide or a conveyance, or to find lodging and food. Her personal situation exacerbated the difficulties of traveling: she was often stuck or lost in a literal and metaphorical sense, suffering from frustrated motion and emotion.

The *published travelogue* begins at sea, but the journey begins on board of the moored ship, as is evident from her *private letters* to Imlay. Mary A. Favret notes that Wollstonecraft frequently longs to escape "oppressive stasis"[23] but does not take notice of the very beginning of the journey before she can leave England. In her letter to Imlay from 13 June 1795, Wollstonecraft promises to do everything to solve her detached lover's business in Sweden and hopes that her efforts would reunite her and her child with him (*Imlay* XLIV, 410). At the same time, Wollstonecraft is afraid of leaving England and losing her lover: "A thousand weak forebodings assault my soul, and the state of my health renders me sensible to every thing. [...] My hand seems unwilling to add adieu!" (XLVII, 412). The body seems to resist the traveler's departure, and so do circumstances. The wind does not yet allow her to depart, and despair and depression make her lose her vitality, responsivity, and agency: she writes from the ship anchored in Hull on 16

June 1795: "I seem to be fading away – perishing beneath a cruel blight, that withers up all my faculties" (XLVII, 412).[24] Her enforced immobility on board ties in with being affected by a paralyzing depression that alienates her from herself. The motion of the ship at anchor wears her down, as do her emotions regarding the relationship she is stuck in:

> It is indeed wearisome to be thus tossed about without going forward. – I have a violent headache – yet I am obliged to take care of the child who is a little tormented by her teeth, because [her maid Marguerite, M.M.] is unable to do anything, she is rendered so sick by the motion of the ship, as we ride at anchor. These are however trifling inconveniences, compared with the anguish of mind – compared with the sinking of a broken heart.
>
> (XLIX, 413)

Instead of setting out and rising to the occasion as intended, the embodied feeling prevails that she is stuck and sinking. At first, her pre-occupied mind renders her insensitive to what otherwise would easily have offended her middle-class sensibility, but five days on board without getting away make her disgusted with "every outward object" and the "disagreeable smells" on board (L, 414). Imprisoned in body, mind, and on board, the beginning of the journey is anything but promising. Chances are that she would have never arrived: "On ship-board, how often as I gazed at the sea, have I longed to bury my troubled bosom in the less troubled deep" (LV, 418). Her repeated yearnings for death testify to her alienation from her lover, the male crew, and herself: she no longer feels at home in her own body. Alienation begins at home.

It seems characteristic for the waywardness of early sea voyaging that, after having finally crossed the channel, the wind drives the ship past two ports, and then a lull renders her immobile again. The rocky coast of Sweden does not allow the ship to land. Her external and internal situations seem like an assault on her embodied, sentient self and her project. The situation compels her to the position of a patient – but without patience. Yet, she finally manages to get ashore by boat.

6.3.2   *The Fall: Losing Her Mind*

Unexpectedly, her alienation from herself and the world in the English harbor even takes a turn for the worse after her arrival in Sweden. No critic I am aware of discusses this momentous accident. On the way to the carriage, she suffers from a fall: "I fell, without any previous warning, senseless on the rocks – and how I escaped with life I can scarcely guess. I was in stupor for a quarter of an hour; the suffusion of blood at last restored me to my senses – the contusion is great, and my brain confused" (*Imlay* LII, 416).

The pathos of her experience cannot be translated into a meaningful answer: being "senseless" does not make sense. No longer feeling her body, she is reduced to a nobody, a radical Other. Her transition from near death to life, passing the threshold between sensing nothing and something, remains inexplicable. The writer appropriates detached, quasi-medical discourse in order to grasp the passage between not me and me. Her "stupor" can only be diagnosed by the self in retrospect, as it is a "state of insensibility or unconsciousness; spec. in Medicine, a condition of near-unconsciousness characterized by great reduction in mental activity and responsiveness".[25] The "restored me" signifies the reflexive sense of the self as it is dependent upon the medically conceived "suffusion" of blood: the self needs to be affected by the senses to make sense of itself. Writing the letter, Wollstonecraft is still affected by the fall but leaves medical discourse behind when she describes her present condition as "confusion" (not the medical 'concussion', which seems to be the 'correct' answer to her injury). Given that the highly educated traveler takes pride in her sharp mind, this must have been a somewhat unpleasant confession.

Instead of rest or medical treatment, the traveler meets with further discomfort: "Twenty miles ride in the rain, after my accident, has sufficiently deranged me – and here I could not get a fire to warm me, or any thing warm to eat; the inns are mere stables [...] I am not well, and yet you see I cannot die" (*Imlay* LII, 416). The suffering traveler, understandably, narrows down her attention to the basics of life. Paradoxically, it seems as if both the accident and its survival lead to self-alienation because her own body eludes her consciousness and her will, expressed in her repeatedly confessed desire to commit suicide.

Why does the writer omit most of her initial experience of the journey from the published travelogue? The conventionally expected excitement at the beginning of a journey turns into its opposite, the unwillingness and inability to depart. The fall as a near-death experience leads to a total loss of control. The unconscious victim of her nerves must have become a helpless object exposed to the gaze of others, at least to that of her timid nurse Marguerite, her one-year-old daughter Fanny, and probably a few strangers.[26] Her own body turns her into a patient, but she represses other responses to her fall, which must have been a disturbing spectacle to her companions and natives. Instead of letting the fall destroy her self-image of the sensitive but capable traveler, she represses the gaze and voices of others and focuses on her own attempts at literally and metaphorically coming to terms with herself, trying to regain mastery. The loving woman may have reported her fall to her lover to invite sympathy, but the public traveler represses this image as a potential metonymy of the woman who is subjected to her body. Thus, the repression of her co-travelers' responses in the published travelogue goes hand in hand with preventing her public readers from seeing her being reduced to a helpless body and strange spectacle. In her private letters, the suffering traveler remains at the mercy of the weather and poor conditions, which focus her attention on herself rather than the foreign country. While the letters to Imlay are often concerned with their complicated relationship that compound the traveler's suffering, the published travelogue tries to find more of a balance between pathos and an adequate response to the foreign landscapes and societies.

6.3.3   *The Falls of Trolhaetta: Searching for Bearings*

Wollstonecraft's experience of the Trolhaetta cascades "extends beyond the conventional reaction of sublime reverie" in "its sense of dislocation and disappointment".[27] Whale is right but does not take account of the traveler's literal and metaphorical searching for her position in and towards nature between pathos, her conventional guidebook answers, and her creative answers. The traveler's sense of dislocation begins with her attempt at assuming a location, connecting motion and emotion. The traveler positions herself as a middle-class, cultivated woman, as an aesthetic observer as opposed to Others with a pragmatic interest. That said, finding the right place for observation turns out to be a challenge for her.

The Trolhaetta Canal (now part of the Göta Canal) was built from 1795 to 1800 to connect "Lake Vänern with Gothenberg and the North Sea" (Wollstonecraft, *A Short Residence* XVII, 316), an enormous project that occupied 900 men at a time and promised a good return on investment. The male workforce exerted their strength to alter nature's course, thereby destroying nature's face. The female traveler does not find the economic endeavor appealing: the work on the canal interferes with her appreciation of nature. The workers' functional motion is contrasted with her "wander[ing] about" (316) in order to obtain a better view on the "grand object" (316) of the cascades, which, however, is also disappointing at first. The observer needs to find a better vantage point to get a pleasant view of the scene.

An island in the Trolhaetta cascades gives a picturesque impression in a landscape of barren, jumbled rocks where chaos and "sterility itself reigned with dreary grandeur" (316):

> [A]t last coming to the conflux of various cataracts, rushing from different falls, struggling with the huge masses of rock, and rebounding from the profound cavities, I immediately retracted, acknowledging that it was indeed a grand object. A little island stood in the midst, covered with firs, which, by dividing the torrent, rendered it more picturesque; one half appearing to issue from a dark cavern, that fancy

might easily imagine a vast fountain, throwing up its waters from the very centre of the earth.

(316)

Her choice of "acknowledging" suggests that she was guided by reports about the grandeur of the scene, but she distinguishes herself as a connoisseur of the sublime and the picturesque, which require an appropriate distance from the scene, and as a Romantic soul, whose mind is stimulated to imagine more than that meets the eye. Her creative answer, the wandering mind of the wanderer, surpasses the set piece of a conventional guidebook. She goes on:

I gazed I know not how long, stunned with the noise; and growing giddy with only looking at the never-ceasing tumultuous motion, I listened, scarcely conscious where I was, when I observed a boy, half obscured by the sparkling foam, fishing under the impending rock on the other side. How he had descended I could not perceive.

(316)

Ironically, in spite of having "retracted" (316) to a safe location, it seems that the noise and motion of the cascades overwhelm not only her senses but also the sense of herself. Pathos drowns consciousness and therefore renders void the attempt to arrive at a literally and metaphorically firm position. Hers is a problem of refined susceptibility that a local boy fishing beneath a rock does not seem to have. However, this is a conjecture suggested by Wollstonecraft's juxtaposition of herself as the observer and the boy supposedly interested in fishing rather than beauty, although she does not know whether staring into the water does not transport him into a daydream or visionary trance, which may well be possible since he does not seem to notice her. Be that as it may, she neither has access to his position and his perspective nor does she try to obtain it, for example, by gesturing to him or joining him. It seems that the boy is in the know and in harmony with nature, a position the traveler is longing and struggling for. This impression is confirmed by the following scene in which she positions herself 'in nature', without, however, having an intuitive grasp 'of nature': she stands on a rock, both above and as close as possible to the object, "a kind of bridge formed by nature, nearly on a level with the commencement of the fall" (317). The "water precipitated itself with intense velocity down a perpendicular, at least fifty or sixty yards, into a gulph, so concealed by foam as to give full play to the fancy: there was continual uproar" (317). In opposition to the scene above (imagining a fountain fed by subterraneous forces), she describes the fall with Latinate words that could have come from a guidebook but does *not* give voice to her fancy stimulated by the assault on her ear, the "continual uproar", and the limitation imposed upon her eyes "by foam". Instead, the traveler reveals her ignorance, in implicit opposition to the boy, about the connection between the "torrent" and "the purling stream" (317), which she needs to deduce from her observation of a drifting log that disappears in the cascade and reappears in the stream.

In sum, the Trolhaetta cascades reveal that the traveler is out of synch with both 'intruders' into nature, the workers on the canal and the boy as a 'child of nature'. The traveler identifies herself as a wanderer on unfamiliar ground. She roughly knows what to see, but her *information* does not seem to prepare her adequately for the *pathos* of her experience. She needs to find her own way of experiencing the overwhelming sight that is enhanced by the overpowering sound, albeit informed by the picturesque and the sublime as *forms* of answering that neither determine the *content* nor exhaust the response to the Other. Relocating and repositioning herself in order to get the most intense experience, Wollstonecraft demonstrates that the traveler abroad is both the agent and patient of dislocation and alienation. In other words, dislocation is not necessarily imposed upon the subject by an alienating society, but in a literal sense the basic condition of traveling and in a metaphorical sense the risk the embodied traveler takes when venturing into the alien.

6.3.4   *Finding Her Place*

Wollstonecraft finds her position as an embodied traveler at the fringe of – but superior to – society and then in the bosom of nature. The traveler offers scenes of suffering from poor conditions abroad and disappointment in love, but counters these weak self-images by the juxtaposition between herself as a lady who exercises body and mind and foreign middle-class women who do neither.[28] She proudly notes in the first letter of her travelogue that the first Swedish woman she meets recognizes her as "*the* lady" (*A Short Residence* I, 246, my emphasis) by looking at her hands, read as a signifier of her superior status. Since she concedes that they can only exchange smiles, not sharing a language, the traveler might only have communicated her own conjecture of the Other's perspective as a confirmation of her own self-image, subordinating – without quite effacing – the Other's response (smile) to her answer (lady). Her positive self-image, which transcends gendered limitations, is also corroborated by the remark of one of her first hosts about her being a "woman of observation" who asks "*men's questions*" (I, 248, emphasis in text).[29]

The middle-class traveler evinces a humane concern for the laboring poor but reveals an ambivalent attitude towards their bodies.[30] However, she also finds fault with women's diets and bodies of her own class, taking herself as the norm. Considering women, she pities servants who earn little and are worked hard, for example, by making them wash linen in the river in winter, which renders "their hands, cut by the ice, [...] cracked and bleeding" (*A Short Residence* III, 253). However, her plea for justice and equality is curtailed by middle-class condescension.[31] Being herself a 36-year-old mother of a small child, who takes pride in her understanding and restraint, she looks down on her timid maid Marguerite and casti-gates "the total want of chastity in the lower class of women" and the "voluptuousness" (*A Short Residence* IV, 258) of the unsophisticated rural youth:

> The country girls of Ireland and Wales equally feel the first impulse of nature, which, restrained in England by fear or delicacy, proves that society is there in a more advanced state. [...] Health and idleness will always account for promiscuous amours; and in some degree I term every person idle, the exercise of whose mind does not bear some pro-portion to that of the body.
>
> (IV, 258)

In the eyes of Wollstonecraft, middle- and upper-class women may be less adventurous than country girls, but also lack mental and physical exertion: "The Swedish ladies exercise neither sufficiently; of course, grow very fat at an early age [...] they are not remarkable for fine forms" (IV, 258).[32] They may have fine complexions, but the over-abundant, very sweet, heav-ily spiced food and drinks as well as the "want of care, almost universally spoil their teeth, which contrast but ill with their ruby lips" (IV, 258).[33] Wollstonecraft castigates the men's indulgence in too much food and drink as well, but does not often comment on their bodily appearance.[34] Once she has to spend the night in an inn overcrowded due to a fair and feels disgusted by the stench that assaults her like a "hot vapour" (XVI, 314) pouring forth from people and animals sleeping in a room like a pigsty, she has to pass in order to get to her own room. She feels like a victim of alien effusions that threaten to contaminate her body. The middle-class traveler literally finds her place apart from and above the abject bodies of the rabble she attributes a lack of "affection or sentiment" to (XVI, 314) and covers the bed with her own clean linen before she goes to sleep, demonstrating her refined sensibility and need of hygiene.[35] Wollstonecraft gives voice to her disturbing experience of Otherness through the aggressive vapor, but her intentional response in separating herself in terms of space and class helps her and her readers to safely contain this strange encounter.[36]

Despite the fact that she often meets with – if not always enjoys – hos-pitality, she feels mainly alienated from the societies of Scandinavia due to their mercenary and sensual priorities rather than her own intellectual, moral, and aesthetic values.[37] Her alienating experience of society does not lead to an alienation from herself, which would be a possible consequence according to Waldenfels. On the contrary, the traveler uses the encounter with foreigners to consolidate her self-confidence because she subjects the Other to her familiar norms, stabilizing the embodied self that was nearly shattered by her alienating experience at the beginning of her journey. The other societies are taken as an opportunity to hold forth her life as a model to others in the framework of the perfectibility of women and society.[38]

Nature often, but not always (see Trolhaetta cascades, above), offers a resource and a refuge from society, whether in the sense of a safe harbor or death by suicide. "Nature is the nurse of sentiment – the true source of taste", she writes, and the observation of nature "excites responsive sympathy", "the harmonized soul" being moved "like the aeolian harp agitated by the changing wind" (*A Short Residence* VI, 271). The senses and the soul, the self and nature seem to form a harmonious bond through responsive affect rather than detached reflection. This model of embodied perception, of course, has been developed through reflection in the discourse of sensibility. The following scene exemplifies her theory: when leisurely wandering, she finds her ideal solitary retreat in a bay near Tønsberg. Instead of taking the position of picturesque distance, the embodied observer places herself in the scene.[39] Nature provides rest and a safe place to slumber, and she awakes "with an eye vaguely curious" to observe the idyllic shore and seascape:

> Every thing seemed to harmonize in tranquility [...]. With what ineffable pleasure have I not gazed – and gazed again, losing my breath through my eyes – my very soul diffused itself in the scene – and seeming to become all senses, gilded in the scarcely-agitated waves, melted in the freshening breeze, or taking its flight with fairy wing, to the misty mountains which bounded the prospect, fancy tript [sic!] over new lawns, more beautiful even than the lovely slopes on the winding shore before me. – I pause, again breathless, to trace, with renewed delight, sentiments which entranced me, when, turning my humid eyes from the expanse below to the vault above, my sight pierced the fleecy clouds.
>
> (*A Short Residence* VIII, 280)

These sensations do not merely affect the self but seem to absorb the self in its senses to the degree of something like a shock to the respiratory system: sensation almost comes too close to home. The writer elaborates how past experience and present recollection are not separated in terms of sense experience and cognitive reflection, but clearly connected to each other through embodied responses. In recollection, the present self is affected by itself in sympathy. 'Higher' cognitive functions are intricately connected to the 'lower' nervous system. This insight ties in well with Whale's[40] and Bohls's[41] finding that Wollstonecraft is daring when she connects this aesthetic experience with a recollection of sexual experience in the subsequent paragraph, coded as a warm heart, a glowing bosom, and blushing. I would add that the link between her aesthetic and sexual susceptibility is "the extreme affection of [her] nature" (VIII, 280), the intricate connection between her senses and heart. However, the temporary location of herself in a liminal – if privileged – position in Scandinavian society and at the heart of nature comes to an end with the journey.

6.3.5 Homeless and Lost

The animating experience of nature abroad stands in stark contrast with the beginning and the ending of her journey, marked by her alienation from home: "The possibilities of homecoming and of national belonging are unavailable to the abandoned and deracinated subject."[42] Wollstonecraft's feeling of loss increases towards the end of her journey and is even more pronounced in her private letters to Imlay. She writes from Copenhagen, just devastated by a great fire, on September 6, 1795: "I see here nothing but heaps of ruins, and only converse with people immersed in trade and sensuality. I am weary of travelling – yet seem to have no home – no resting place to look to. – I am strangely cast off" (*Imlay* LXV, 426–27). The destruction of homes in the great fire seems to resonate with her own feeling of being homeless: the observed place forms the objective correlative of the embodied observer. Comparable to the beginning of her journey, she feels suspended, uncertain at what time she "will have landed, or be hovering on the British coast" (LXVII, 428), and whether Imlay will meet her or not. However, by trying to protect herself from her unfaithful lover, she contributes to her liminal position by threatening to have an acquaintance procure her "an obscure lodging, and not to inform any body" of her arrival (LXVII, 428–29).

Nearing the close of her journey in the travelogue, it seems that England gradually loses its function as the norm and gauge of foreign countries. For example, she notices the similarity of the rich oppressing and disdaining the lower classes across nations (*A Short Residence* XXII, 338; XXIII, 339). While praising the service of English pilots in her first letter, she despises the "cunning servility" of English innkeepers and waiters in one of the last letters (XXII, 338). To the weary traveler, whose experience of foreign nature has changed expectations, the iconic cliffs of Dover do not appear quite so enchanting as before: "at the sight of Dover cliffs, I wondered how any body could term them grand; they appear insignificant to me, after those I had seen in Sweden and Norway" (XXV, 345). Rather than being enriched by her experience, she seems to be more alienated than before. The Other mostly cherished as a negative counter-image of the progressive traveler and her mother country England seems to vanish with the prospect of a home that does not fulfill its promise of a safe haven. In addition, her "spirit of observation seems to be fled" (XXV, 345), as if the vibrant sensibility she was full of in Scandinavian nature had deflated. Dreading to go home without having a home, the journey takes her back full circle to the position of a liminal self, alike alienated from herself and society. Her repeatedly uttered desire to put an end to herself, and her attempt at committing suicide briefly after her return to England can be understood as an escape from an unbearable situation and the expression of a self that is radically alienated both from her body and from society.

6.4 Conclusion

Alienation begins at home and is intensified on the threshold between home and abroad. The private letters to Imlay about the delayed begin-ning of the journey and its discontent, which cover the days before the public travelogue begins, foreground the embodied traveler's plights, who is neither able nor willing to set out. Her suffering from an impasse in her relationship corresponds to her frustration due to her arrested mobil-ity on board of her ship riding at anchor. She is sick of life on land and on board, considering suicide. She is alienated from her body as it eludes her consciousness in a nearly fatal accident that leaves her unconscious. Furthermore, she is alienated from her will in being resilient to the adver-sity of traveling in spite of her desire to die rather than continue living a life of suffering. The nearly fatal accident, omitted from the public let-ters, reveals the gap between pathos and response in the laborious attempt to come to terms with a truly disconcerting experience. It seems that the traveler is caught in a vicious circle of alienation from her lover, society, and herself, suffering from a depression that wears her down and often colors her experience. Traveling requires an extra bodily and mental effort to position oneself and move in an extraordinary society and territory. It is ironic and paradoxical that she profits from the alienating societies of Scandinavia in reconfiguring her impaired self as superior in bodily health and taste primarily in contrast to women, and in intellect and morals in contrast to men. This form of empowerment is only possible because she tries to contain the pathos of alienating experience in developing answers within the framework of enlightened perfectibility. Her encounters with nature abroad are ambivalent, but she finds creative answers to her new experience: while she is still searching for a position towards nature at the Trolhaetta cascades, she finds something of an embodied communion with nature at a bay near Tønsberg. In conclusion, the traveler needs the Other to reconfirm her identity as an English middle-class reformer with a refined aesthetic sensibility and sense of nature. Ironically, her experience abroad increases her alienation from English nature and culture upon her return, which fail to live up to her hopes. In sum: Wollstonecraft's liminality and alienation begin and end at home.

Notes

1 Mary Wollstonecraft, *Letters Written during a Short Residence in Sweden, Norway, and Denmark. The Works of Mary Wollstonecraft* 6, eds. Janet Todd and Marilyn Butler (London: William Pickering, 1989), pp. 237–348. Further references to this edition are included parenthetically in the text. Roman numbers indicating the letters precede the Arabic page numbers.

2 See, for example, Karen R. Lawrence, *Penelope Voyages: Women and Travel in the British Literary Tradition* (Ithaca: Cornell University Press, 1994); Karen Hust, "In Suspect Terrain: Mary Wollstonecraft Confronts Mother Nature in *Letters Written during a Short Residence in Sweden, Norway, and Denmark*", *Women's Studies*, 25 (1996): 483–505; Stephanie Buus, "Bound for Scandinavia: Mary Wollstonecraft's Promethean Journey in 'Letters Written during a Short Residence in Sweden, Norway, and Denmark'", *Scandinavica*, 40, no. 2 (2001): 241–61; Mary A. Favret, "*Letters Written during a Short Residence in Sweden, Norway, and Denmark*: Traveling with Mary Wollstonecraft", in Claudia L. Johnson, ed., *The Cambridge Companion to Mary Wollstonecraft* (Cambridge: Cambridge University Press, 2002), pp. 209–27; Anne S. Sørensen, "Mary Wollstonecraft's Politics of the Picturesque", in Anka Ryall and Catherine Sandbach-Dahlström, eds., *Mary Wollstonecraft's Journey to Scandinavia: Essays* (Stockholm: Almqvist & Wiksell, 2003), pp. 93–114; Christine Chaney, "The Rhetorical Strategies of 'Tumultuous Emotions': Wollstonecraft's Letters Written in Sweden", *Journal of Narrative Theory*, 34, no. 3 (2004): 277–303; and Michael Meyer, "Romantic Travel Books", in Ralf Haekel, ed., *Handbook of British Romanticism* (Berlin: De Gruyter, 2017), pp. 237–55.

3 Elizabeth A. Bohls, *Women Travel Writers and the Language of Aesthetics, 1716–1818* (Cambridge: Cambridge University Press, 1995), pp. 141–42, 146, 165–67.

4 Bohls, *Women Travel Writers*, p. 160.

5 Bohls, *Women Travel Writers*, p. 155.

6 John Whale, "Death in the Face of Nature – Self, Society and Body in Wollstonecraft's 'Letters Written in Sweden, Norway, and Denmark'", *Romanticism*, 1, no. 2 (1995): 177–92 (at pp. 177–79).

7 Mary Wollstonecraft, *Letters to Imlay*, in Todd and Butler, eds., *The Works of Mary Wollstonecraft* 6, pp. 365–438. Further references to this edition are included parenthetically in the text. Roman numbers indicating the letters precede the Arabic page numbers. Where necessary for clarity, these private letters are marked with the short title *Imlay* as opposed to Wollstonecraft's travelogue with the short title *A Short Residence*.

8 Carl Thompson, *The Suffering Traveller and the Romantic Imagination* (Oxford: Oxford University Press, 2007), pp. 6–8, 171.

9 Bernhard Waldenfels, *Phenomenology of the Alien: Basic Concepts*, trans. Alexander Kozin and Tanja Stähler (Evanston: Northwestern University Press, 2011).

10 Graham J. Barker-Benfield, "Sensibility", in Iain McCalman et al., eds., *An Oxford Companion to the Romantic Age. British Culture, 1776–1832* (Oxford: Oxford University Press, 2001), pp. 102–14.

11 Waldenfels, *Phenomenology*, p. 75.

12 Waldenfels, *Phenomenology*, p. 16.
13 Waldenfels, *Phenomenology*, p. 77.
14 Waldenfels, *Phenomenology*, p. 53–55.
15 Waldenfels, *Phenomenology*, p. 28.
16 Waldenfels, *Phenomenology*, p. 37, emphasis in text.
17 Waldenfels, *Phenomenology*, p. 37.
18 Waldenfels, *Phenomenology*, p. 36, emphasis in text.
19 Waldenfels, *Phenomenology*, p. 41–42.
20 Waldenfels, *Phenomenology*, p. 32, 41.
21 Waldenfels, *Phenomenology*, p. 3.
22 Compare Favret, "*Letters*", p. 209; and Chaney, "Rhetorical Strategies of 'Tumultuous Emotions'", pp. 291–92.
23 Favret, "*Letters*", p. 214.
24 Compare *Imlay* LV, 419.
25 MobiSystems, Shorter English Dictionary App. "Stupor", *MobiSystems, vers. 11.4. Google PlayStore, https://play.google.com/store/apps/details?id=com.mobisystems.msdict.embedded.wireless.oxford.shorterenglish&hl=en_US&gl=US [18 November 2022].
26 The relationship between mother and child forms the topic of Elizabeth Zold's contribution to this volume and is therefore slighted here.
27 Whale, "Death in the Face of Nature", p. 182.
28 See, for example, Wollstonecraft, *A Short Residence*, VIII, 280–81 and LXI, 423.
29 Anthony Pollock, "Aesthetic Economies of Immasculation – Capitalism and Gender in Wollstonecraft's Letters from Sweden", *The Eighteenth Century: Theory and Interpretation*, 52, no. 2 (2011): 193–211 (at p. 196).
30 Bohls, *Women Travel Writers*, pp. 157–58.
31 Bohls, *Women Travel Writers*, p. 156.
32 Compare Whale, "Death in the Face of Nature", p. 183.
33 For a counterexample of a pretty young woman in Germany, see *A Short Residence* XXII, 338.
34 See, for example, *A Short Residence* XXIV, 343.
35 Bohls, *Women Travel Writers*, p. 157.
36 See also Maruo-Schröder's article in this collection, in which she discusses a similar form of othering the lower-class body by the middle-class traveler.
37 Pollock, "Aesthetic Economies of Immasculation", p. 198.
38 Chaney, "Rhetorical Strategies of 'Tumultuous Emotions'", pp. 295–96.
39 Bohls, *Women Travel Writers*, p. 162.
40 "Death in the Face of Nature", pp. 183–84.
41 *Women Travel Writers*, pp. 163–64.
42 Katherine Turner, *British Travel Writers in Europe 1750–1800: Authorship, Gender, and National Identity* (Farnham: Ashgate, 2001), p. 237. Compare Lawrence, *Penelope Voyages*, pp. 82–84.

7 "The Most Dirtiest Children"

Spectacles of Otherness on the American Frontier

Nicole Maruo-Schröder

7.1 Introduction: Sights and Sceneries in Travel Writing

Sights – the descriptions of landscapes, sceneries, and places worth see-ing – are a staple element in travel writing and serve as much to underline the traveler's (and writer's) knowledge and authority as they are meant to entice potential travelers (and readers) to go out and see them. "Seeing and being seen are the conventional indicators of cultural capital for the traveler"[1] as Margaret Topping rightly points out. It is therefore not sur-prising that sights, which range from bucolic and pastoral sceneries to the most spectacular and sublime landscapes, can also be found in women's travel writing on the American frontier, a geographical, political, and mythical region that has inspired the American imagination ever since its inception. Despite hardships and even dangers to be faced while trave-ling there – and much in contrast to the stereotypical image of the help-less and fearful woman traveler – female writers frequently celebrate the frontier in their texts, not just in terms of its spectacular landscapes but also with regard to the freedom and independence they experience there. Susan Shelby Magoffin, for instance, who traveled to the frontier in June 1846 accompanying her husband, a seasoned trader, writes in her diary:

> Oh, this is a life I would not exchange for a good deal! There is such independence, so much free uncontaminated air, which impregnates the mind, the feelings, nay every thought, with purity. I breathe free without that oppression and uneasiness felt in the gossiping circles of a settled home.[2]

Much in contrast to stereotypical ideas of an exhausted, displaced, and endangered (female) body on the frontier, the passage positively highlights independence and freedom from social constraints, an idea that is under-lined by the implied contrast to the 'contaminated' nature of home.[3] Such a celebration reclaims the mythic image of a pure and invigorating frontier space from a woman's point of view. Moreover, Magoffin's image of breath-ing free seems to question the domestic values of home and the regulations

DOI: 10.4324/9781003331803-9

they come with, highlighting that such constraints are limiting, which is surprising since women were considered to be the arbiters of domesticity. In opposition to the "settled home", the experience of the landscape is cathartic, purifying, and invigorating her mind as well as her body.

Magoffin's comment is not the only one that shows an appreciation of the frontier as a space that affords freedom. Women ride, walk, and move around, sometimes daunted by the wilderness, frequently inspired. Eliza Wood Farnham's travel account *Life on the Prairies* (1846) is another case in point. Traveling the Illinois frontier in the 1830s, she visits her sister, who says that nature provides "[f]reedom from want, [...]; freedom from social trammels; freedom from the struggles of an emulation founded in vanity or other vitiated desires; from the myriad forms of ruinous and slavish excess [...]".[4]

In contrast to nature, social regulations here appear unnatural and unnecessary, creating a body hampered by desires and feelings that were often felt to be artificial, imposed by social regulations and practices.[5] Indeed, in addition to the freedom from such regulations, nature also provides a space for the spiritual, transcendental experience of the divine (see Farnham 44–45, 130). Similarly, Isabella Bird, on her famous journey alone through the Rocky Mountain area in 1873, enjoys her travels on horseback, where she can take pleasure in "a cool atmosphere of exquisite purity",[6] ride in her "own fashion" (10), and "feas[t] [her] eyes on pines" (12) after her experience of the "very repulsive city of Sacramento" (4). Even Sarah Royce, who accompanied her husband to California on a rather dangerous and strenuous journey in 1849, finds nature to be a spiritual place where she feels God's presence, even and especially amongst danger and peril.[7] And, finally, in her account *A New Home, Who'll Follow?* (1839) about life on the Michigan frontier in the 1830s, Caroline Kirkland finds the frontier anything but dangerous. In fact, in her highly readable account she makes fun of fictional and factual narratives that characterize the frontier as a place for dangerous adventures and masculine heroics: "'Tis true there are but meager materials for anything which might be called a story. I have never seen a cougar – nor been bitten by a rattlesnake."[8] To her, one of the most 'perilous' aspects of frontier life, never even mentioned in the frontier adventures written by men, relates, in fact, to domestic matters: "The inexorable dinner hour, which is passed *sub silentio* in imaginary forests, always recurs, in real woods, with distressing iteration, once in twenty-four hours, as I found to my cost" (49). Facing dinner time without a proper kitchen needs courage and strength, and preparing family meals is about as strenuous as it gets (for her) on the Michigan frontier.

Yet, as invigorated and inspired as these travelers seem to have felt facing the freedom of the frontier wilderness, as apprehensive, even anxious they appear when they encounter others, specifically lower-class settlers, enjoying the very same freedom. In what now follows, I focus on travel narratives by Eliza Farnham, Caroline Kirkland, Susan Shelby Magoffin, and Susanna Moodie, showing that what is at issue in these texts is not so much the domestication of the frontier wilderness as such, but, more precisely, the regulation of the bodies who inhabit it. Hence, I argue that, in addition to the views and vistas of nature sceneries, the travel writers stage what one could call spectacles of otherness, a narrative display of the bodies of 'others' that seeks to impress upon readers the dangers that freedom can have for bodies that need disciplining. In the eyes of the travelers discussed here, this lack of discipline is visibly etched onto these bodies as dirt and ugliness, even animal-like features. Although the women writers themselves sometimes struggle with frontier life and the limitations it imposes upon, for instance, body-related housekeeping practices, this is never portrayed as a dangerous deviation. On the contrary, it seems to contain the promise of a better future. It is the bodies of those who are seen as different – in terms of education, financial resources, lifestyle, etc. – that need to be regulated. The 'other' is, of course, only a relational construction from the point of view of the perceiver so that in the travel accounts a variety of 'other bodies' is encountered and described from the perspective of the travel writer. To remain within the constraints of the essay, I focus on the class-based perception of otherness, leaving out, for instance, the construction of otherness based on gender or race.[9] Such deviant bodies threaten to 'contaminate' the frontier, sometimes even the writers themselves, a contamination that is, as I discuss below, expressed by the disgust the travelers feel at their sight – or smell or touch. Moreover, such 'de-civilized' bodies threaten to 'breed' deviation and immorality – metaphorically but also literally – if they are left unchecked, or so the narratives suggest, presenting to their readers' scenes that expose the lower-class body as dangerously undisciplined. These encounters with and descriptions of otherness function as 'unseemly sights' in the narrative, as spectacles even, which spoil the narrators' travel experience of the frontier and, significantly, are indicative of a threat not just to the West but to the whole nation. I choose the concept of 'sight' deliberately, as it highlights the role that the sense of vision and its implied values of objectivity and empiricism play in such encounters; it also links up with ideas of control and domination, expressed in postcolonial concepts such as Mary Louise Pratt's "imperial eye"[10] (1992) with which travel writing has been critically analyzed.

Considered as free, open, and 'empty', the frontier served as a space of projection for fantasies about the future of the nation; it appeared as a space of boundless opportunities. Yet, in the eyes of the travel writers, it needed domestication and disciplining in so far as it threatened to produce unregulated, de-civilized bodies, vividly described and criticized in the travel accounts I analyze below. In the following, I briefly discuss the notion of the frontier as a mythic space of freedom and renewal as well as danger and deviance (as part of the discourses surrounding national identity at the time), before I turn to the concept of embodied traveling, perception, and the gaze. The main part of the paper will then analyze selected scenes from travel accounts in which the other body is presented as a spectacle of deviance. More specifically, I focus on strategies of looking and the ways in which the travelers' own body experience as well as the travelees' bodies[11] are represented, showing how the various sense perceptions govern the relation between traveler and travelee, and between self and other in sometimes contradictory ways.

7.2   Traveling Bodies on the Frontier: Perception and Representation of Self and Other

Originally coined by historian Frederick Jackson Turner (1893) in his famous claim that at the end of the nineteenth century the frontier was closed, the term is less a geographically specific designation than a myth that reverberates with other images and ideologies circulating in American culture until today, among them freedom, individuality, and boundless opportunities as well as a certain notion of masculinity.[12] Historically, the frontier designated the regions that were about to be settled, in Turner's words, and in nineteenth-century perception, "the meeting point between savagery and civilization".[13] As such, it was a place of transformation and transition, both temporally and geographically. Its very conception as unsettled, 'empty' territory served to justify colonization, including the removal of Native American tribes; moreover, and connected to this, the idea of manifest destiny suggested that this colonization was as necessary as it was inevitable. The frontier was seen as an 'unused' and unregulated territory, a promised land which fueled dreams about endless opportunities while also threatening violence and disorder. It was a space for projections concerning both the origins and the future of the American nation. The frontier, in other words, contained the promise of new beginnings but was also a place that lacked culture, order, and infrastructure and thus had to be conquered and domesticated.

This frontier myth was very much a part of the cultural imagination of the time, ranging from sociopolitical discourses to (popular) cultural texts as diverse as adventure novels and landscape paintings. While much of this popular imagination was intertwined with discourses of (violent) masculinity,[14] women were an active part of the production of such imaginations and images about the frontier as Annette Kolodny has shown.[15] These images were, in the words of Brigitte Georgi-Findlay, "at times less alternative than complementary to the Adamic myth"[16] central to the male imagination. Both men and women took part in the discourses as well as the activities surrounding the settling of the West, albeit in different ways.

Particularly the idea of domestication linked up closely with gendered ideologies of the time, such as the image of the "Republican Mother", or the ideologies of "true womanhood" and "separate spheres", which held that women were responsible for creating a home (and by extension a nation) that served to imbue its inhabitants (husband, children, servants) with the right moral mindset.[17] Paradoxically, it was the 'female' domestic sphere – ostensibly separate and different from the 'male' public, political sphere – that was thought to be essential for the formation of a strong and healthy nation by raising appropriate future citizens. Therefore, the idea of domesticating the West went beyond the implementation of economic and political infrastructure and included the creation of homes that ensured the raising of future generations and, thus, a healthy American nation. Likewise important was the Jeffersonian idea of the USA as an agrarian nation, in which land-owning, self-sufficient farmers made up the ideal(ized) citizenry. It is a discourse that becomes visible in a variety of ways in American travel writing, ranging from the western wilderness imagined as a new Garden Eden to the tension between an uncontaminated, free West and a degenerating, urbanized East, visible, for instance, in Magoffin's comment quoted earlier.

As I suggested in my introduction, the traveler's encounter with new spaces and places (and the people who are already there) is governed by the negotiation of the unfamiliar on the basis of what we already know and what we are used to. In this sense, our cultural baggage travels with us, providing not just the framework for an evaluation of our experience but the very grounds which allow us to make that experience in the first place. Perception – our sensual and cognitive 'access' to and interaction with the world – is not neutral and independent of the world around us (biological processes of seeing, hearing, etc., notwithstanding). As Alva Noë points out, consciousness is "something we do, [...] a kind of living activity"[18] and "requires the joint operation of brain, body, and world",[19] which means that perception is always already enmeshed with our environment as well as culturally specific norms, values, and knowledge.[20] We travel as embodied beings, not as independent minds in somehow separate bodies. If attitudes are, in Lisa Heldke's phrase, "embodiments of cultural ideologies",[21] our travel experiences are indicative not just of the places we encounter but also of those we happen to come from.[22] This notion of the 'lived body' makes it all the more interesting to look at the role that bodies and bodily experience play in travel writing. Hence, the travelers' experience of bodies (their own and the others') are not simply descriptions of what they felt and saw on the frontier but are entangled in complex ways with who they were and where they came from, including, e.g., the middle- and upper-class culture they grew up in as well as their expectations about traveling and the frontier itself.[23]

According to Norbert Elias, bodies are always and everywhere regulated and thus shaped by social conventions and requirements, and it is this ongoing regulation that the concept of the "civilized body" refers to. Hence, civilization is not a teleological goal to be achieved, but an ongoing process[24] in which bodies are, as Chris Shilling puts it, "unfinished entities that develop in social contexts, are mutually interdependent and possess emergent properties which are nevertheless subject to processes of flux and change".[25] The way we dress or behave, for instance, does not develop independently, individually, or simply as a 'natural' consequence of our (individual) biological needs, but as part of what Elias calls "figurations", networks "formed by large numbers of interdependent human beings"[26] who are neither completely autonomous from each other nor passive subjects governed "by abstract social facts".[27] It is, in other words, a process in which biological, cultural, and social factors interact.

Elias bases his conception on observations regarding changes in both sociopolitical structures and body conceptions in human history, comparing medieval times to seventeenth-century French court society.[28] Changes in political structure and social practices and values led to changes in bodily behavior, a "progressive *socialization* of the body"[29] that involved an increasing (self) control of its natural functions and emotions, which also led to a compartmentalization of body practices.[30] Many of such bodily practices have become naturalized, i.e., they appear as a natural part of human characteristics and behavior and not as culturally specific. Therefore, the 'civilized body' as it has emerged in these processes has come to be perceived – in appearance, behavior, feelings – in contrast to the 'animal' body. Bodily practices and habits, in other words, can function as a sign of distinction and as such are used to signal refinement and class membership. Accordingly, their presence or absence can then be used to measure the degree of 'humanity', something which becomes visible in (Western) travel accounts in which the narrator's own culture serves as a foil against which she experiences and judges frontier life.[31]

A central element of such evaluations in travel writing is the sense of vision, the gaze. Although in travel literature looking at something or someone is traditionally framed by values of scientific objectivity and truth, the practice of looking is anything but neutral. As John Berger has argued, looking at something inevitably constructs the seen in relation to ourselves: "We never look at just one thing; we are always looking at the relation between things and ourselves. Our vision is continually active, [...] constituting what is present to us as we are."[32] Among other things, the gaze is structured by relations of power, more precisely by a power imbalance in which the "bearer" of the look is active, dominant, and controlling while the person being looked at is turned into a passive spectacle.[33] According to Pratt, just such issues of control and domination are at stake in European travel and exploration writings, in which the traveler-explorer (usually white, male, European) looks at the land and takes possession of it. Although this is done in a seemingly passive and innocent manner (as part of what Pratt calls the strategy of "anti-conquest"[34]), this gaze at the landscape and its inhabitants, coming more often than not from an elevated position, is vital for the production of the colonizer's hegemonic position, a "seeing-man [...] whose imperial eyes passively look out and possess".[35] Measuring and ordering the unfamiliar, the seemingly detached and objective gaze thus helps 'to know' the other, i.e., it is part of the production of knowledge about the colonial subject, a production that – by looking (down) at the other – literally puts the colonized other in their place. In the words of Giorgia Alù and Sarah Patricia Hill, "[i]n travel writing, views and gazes express a narrative space from which narrator and reader scrutinise, judge and categorise the varied cultures and societies they explore through writing and reading".[36] As I argue below, such strategies of looking can also be found in travel writing on the frontier, particularly with regard to 'domestic landscapes' and the others that are encountered there. Similar to the views of natural landscapes that the writers portray, they represent the others as sights to gaze at in wonder, even in shock. This ambiguous moment when the pleasure of looking merges with a certain repugnance at the sight is expressed by the notion of the spectacle here, which – in its common-sense usage – combines ideas of the unusual, the dramatically exaggerated as well as the feeling of contempt; an ambiguity that well captures the anxiety that the travelers claim to have felt in view of what they saw as the unregulated lower-class body on the frontier. As Andreas Niehaus and Uta Schaffers point out in their discussion of the transgressive nature of the gaze, a spectacle was traditionally understood as a performance for the common (i.e., 'lower-class') people, an amusement fit for the less refined;[37] yet here, the spectacle is not so much *for* than *about* these people, a sight which captured above all the anxieties of the middle classes with regard to the colonization of the West and the future development of the nation.

The various discourses about the West and its colonization were a vital part of American travel writing, providing a framework for the expectations about and experience made during actual travel on the frontier, as well as for the ideas and conventions surrounding the genre of frontier (travel) writing. In other words, not even private travel accounts (such as diaries) provide an unfiltered insight into what life and people on the frontier were 'really' like, as these accounts are transformations, not of an unmediated experience but of memories of an experience always shaped by a variety of expectations.[38] With the exception of the diary by Magoffin, the texts discussed here were, moreover, written for publication, so that we can surmise that they are shaped to an even greater degree by genre conventions and audience expectations. Additionally, it is important to keep in mind that the authors were highly literate and educated, coming from the middle and upper classes. While they may not have been wealthy, they had probably more resources at their disposal than the average 'first wave' settler on the frontier. Yet, while certainly not objective accounts of frontier life, these writers' texts can be analyzed as part of the (national) discourses surrounding the future of the West and its colonization, indicative of ideas and ideologies – and 'knowledge' about the West – that are influenced by expectations based on frameworks of culture, class, gender, and race. Given the fact that the domestic sphere was considered to be the 'natural' expertise of women, it is not surprising that – in contrast to texts by male authors – scenes of domesticity and the different bodies inhabiting them are featured quite frequently in these women's travel accounts. Such a concern with the domestic would not just allow female writers to comment on politically pertinent questions and propagate norms and values of the white middle classes, but also give them the authority to do so.

7.3   Unseemly Sights: Other(ed) Bodies in Women's Frontier Writing

> I was pondering on this proffer [of help with domestic chores], when the sallow damsel arose from her seat, took a short pipe from her bosom, (not "Pan's reedy pipe," reader) filled it with tobacco, which she carried in her "work-pocket," and reseating herself, began to smoke with the greatest gusto, turning ever and anon to spit at the hearth. Incredible again? alas, would it were not true!
>
> (Kirkland 56)

There are a number of ways in which bodies – both the traveler's and those she encounters – come to the fore in travel writing, ranging from scenes in which the travelers are pestered by bed bugs and mosquitoes, comments on food and drink (or the lack thereof) to complaints about the climate (heat, cold, rain) and descriptions of nature that do not merely highlight the senses, e.g., looking or listening, but frequently point to aesthetic, even spiritual experiences. My focus here, however, lies on passages such as the one recounted by Kirkland above, which draw attention to what one could call with Elias the 'de-civilized' body; such a body can emerge in the context of the frontier because established (embodied, habituated) rules and regulations are here very much open to negotiation. While the female narrators experience the lack of etiquette, habits, and customs for themselves frequently in positive ways, often using it as a way to criticize society and its overly 'artificial' rules and norms, they are also afraid of what such a lack of control does to 'other bodies' – Kirkland does certainly not approve of spitting and smoking in her home, even less when it is done by a woman. Moreover, in this scene, any absolute differentiation between self and other is complicated: the odor of the smoke marks the woman as different while the fact that she smokes in Kirkland's house makes the smell part of the narrator's home.[39] While the travel writers meet and describe a number of bodies marked – in their eyes – differently by intersections of race, class, and gender, I focus my discussion in the following on the portrayal of the lower-class body in contrast to the traveler's own, 'civilized' middle-class one, a portrayal of difference that is at the same time challenged in a number of ways.

A case in point is Farnham's account of a domestic sphere typical for such unregulated lower-class bodies, a scene in which several of these aspects become visible:

> The house was one of the meanest description of cabins. It [...] showed only a four-light window, or rather sash; for soon after I first saw it, the third was broken out [...]. A patchwork quilt of blue jeans and red flannel was hung across the aperture a few days after, and never removed while I remained in the country. Directly beneath this, [...] was a green pool [...]. It [...] redounded not a little to the taste of some eight or ten large swine, who delighted their senses in its aromatic depths, at the same time that they regaled those of by-passers.
>
> (37f.)

This is not a modest frontier cabin, in tune with the idealized Jeffersonian conception of a simple but dignified agrarian lifestyle, but a dilapidated one, in a state of (permanent) disrepair, both inside and out. The cabin's state of disrepair serves to signify – particularly in the context of nineteenth-century domesticity and motherhood – the degeneration of its inhabitants, and not surprisingly, it houses a family with many children, whose "minimum [number] was eight, the maximum double that number. I rarely saw less than the former, sporting away the morning of life, in their rags and filth" (38). Quite stereotypically, the large number of children goes together with the fact that they are dirty, with stiff hair and stained clothes, playing in and around the pigsty (38). This pigsty seems to be part of the house, suggesting that there is no definite border between humans and (dirty) animals: the pigs are held directly beneath the (permanently) broken window through which the (pigs') dirt and smell find their way into the house, which then becomes permanently polluted by it. While the pigs love the dirt and the smell of the place (as do, by implication, the inhabitants), Farnham emphasizes that their bad smell virtually 'attacks' people who walk by. It is not surprising that the dirty bodies, living in such close proximity to a symbolically dirty animal, cross, in Farnham's view, even the line to the 'subhuman': she perceives the children as "a herd of wild animals" (38), a herd that, by its sheer number, poses a threat, a multiplication of unruly bodies that threaten to overrun the frontier.

In another windowless cabin, furnished, among other things, with a "broken table" and a "dilapidated chair" (39), the narrator encounters a similar view of a degenerated home. While the mother is sitting, smoking a pipe, and picking the foul kernels from her coffee beans only to throw them on the floor, a "child was playing in the ashes" of the fireplace, and the narrator points out, "I have never seen more utter poverty or filth" (39). Not only do poverty and filth here become practically exchangeable, they are also naturalized as part of the settlers' character: this class of people, according to Farnham, shows "the most *disgusting* indifference to the common comforts of a more civilized condition" (40, my emphasis); their unclean and unkempt bodies clearly display this lack of civilization, and, what is more, they are turned into a 'sight' to look at for the reader.[40] The frontier enables a lack of regulation that turns lower-class bodies quite literally into human trash in the eyes of the narrator, a view that is 'breathtaking' in more than one sense. The look that creates the sight of these other bodies is traditionally the sense that stands for objectivity and truth and can consequently be read as a (narrative) tool with which the narrator claims authority; at the same time, the look creates her distance to this scene, emphasizing her difference from these problematic settlers.

However, as Farnham's sensation of disgust suggests, such 'trash' is also polluting and contagious. According to Paul Rozin and April Fallon, disgusting objects "have offensive properties" and, moreover, "the capacity to contaminate",[41] something which becomes visible in the travel accounts as well. Clearly, Farnham's attempt to distance herself from these uncivilized bodies (by way of sight) is undercut by the sensation of disgust which threatens to overcome this distance: just like the pigs' offensive smell attacks the traveler in the first scene, the filth in this cabin (and all its implications for and imaginations on the part of the traveler) threaten to penetrate and contaminate her body. Feeling disgust at the sight of 'other' bodies, then, implies more than a momentous discomfort on the part of the traveler, elicited by filth and poverty; it seems to reveal the deep-lying anxiety that such poverty, filth, and dilapidation are likely to spread like a disease, threatening to turn the frontier from a space of renewal and invigoration into a breeding ground for the downfall of the nation.

Another connected image of contamination can be found in Susanna Moodie's famous account of her life on the Canadian frontier in the 1830s, *Roughing It in the Bush* (1852).[42] Here, too, similar scenes of dirty, animal-like, sometimes simply ugly bodies, filthy clothing, and untidy homes can be found. Describing a house that her husband bought from a settler to save the work of having to erect it himself, the narrator, too, combines images of filth and animality in her description of the settlers' otherness. Moreover, not only is the house "more filthy than a pig-sty", it is also infested with vermin:

> The first night we slept in the new house, a demon of unrest had taken possession of it in the shape of a countless swarm of mice. [...] They scampered over our pillows and jumped upon our faces, squeaking [...]. How Uncle Joe's family could have allowed such a nuisance to exist astonished me; to sleep with these creatures continually running over us was impossible; and they were not the only evils in the shape of vermin we had to contend with. The old logs which composed the walls of the house were full of bugs and large black ants; and the place, owing to the number of dogs that always had slept under the beds with the children, was infested with fleas.[43]

The passage suggests that even after the former inhabitants have left the house, traces of their lifestyle in the form of various kinds of vermin still linger and contaminate the house. The way in which the scene is narratively rendered belies the fact that bugs, insects, and the like were naturally a major part (and problem) of settling in the 'wilderness'. Instead, it serves here to dehumanize the cabin's former inhabitants. Vermin do not just suggest a lack of hygiene but, again, serve as a sign of the settlers' animal-like nature and lifestyle as mice and fleas are typically found on and around animals.[44] The severity of the infestation is underlined by the exaggerated description of the "countless" mice as well as bugs, ants, and fleas with which the place is crawling; the house is 'possessed' by a "demon" that threatens not just to contaminate but to literally 'eat' Moodie and her family so that nothing short of an exorcism can render it safe and inhabitable. This felt threat of contamination is also reflected in the sense impressions that dominate the scene: while earlier passages were characterized by a more distancing look at the settlers, in this scene other senses dominate as the vermin are heard and felt, rather than seen in the middle of the night, something which threatens disorder. As Sarah Jackson points out: "The sense of touch, in particular, is traditionally associated with a proximity to the body and has thus been conceptualised as opposed to order and cognition."[45] The night in the cabin threatens to undermine the (narratively constructed) distance/difference between Moodie and her neighbors that she so frequently insists upon; Moodie's body here becomes porous, penetrated by noise, touch, and bites as the distinction from 'the other', so carefully constructed and maintained by the look, collapses in this scene.

7.4   Looking Back: Transgressions of the 'Other'

While the vermin attack is a rather obvious form of transgression, very much 'felt' by the narrator, other moments that question the difference between the Moodies and their neighbors are not as obvious. In an earlier passage, the narrator points to another aspect indicating the lack of boundaries and control of such settlers: they keep intruding into her own private space, both metaphorically with their questions and literally by entering her house. The neighbors' girls

> would come in without the least ceremony, and young as they were, ask me a thousand impertinent questions; and when I civilly requested them to leave the room, they would range themselves upon the doorstep, watching my motions, with their black eyes gleaming upon me through their tangled, uncombed locks. [...] Their visits were not visits of love, but of mere idle curiosity, not unmingled with malicious hatred.
>
> (Moodie 91)

Again, the stereotypical markers of the lower class become apparent: dirty, unkempt bodies by whose transgressive behavior the narrator feels threatened. Moreover, the children transgress into her life and home with unannounced visits as much as with their questions, and when the narrator tries to hold them in check by removing them from her house, they linger on the threshold and return her scrutinizing and evaluating gaze with "eyes gleaming upon me" (91). The girls' gaze questions and deconstructs the difference, the superiority that Moodie carefully maintains in her account: not only are the children felt to be in the room although they are standing at the door, their gaze also forces the narrator to reflect upon her own self. She feels watched and evaluated in her own home, "obliged [...] to put a painful restraint upon the thoughtfulness in which it was so delightful to me to indulge" (91; see also 62, 113). Such a scene is particularly telling because the traveler's observing and evaluating gaze is thrown back upon itself, and we get at least a fleeting glimpse of the other's perspective. For instance, the "malicious hatred" that is communicated here via the girls' gaze serves as a reminder of the uneven power relations on the frontier, which is anything but a space of true equality. What remains unsaid in this passage but becomes clear from the context is the fact that the Moodies have moved into the family's house because the latter are too poor to pay the mortgage. Having to give up their home, the family naturally feels some resentment toward the new owners and their occupation of this home, which must feel, after all, also like a transgression, even a violation of boundaries, sanctioned though it is by economic power relations.

The gaze is a transgression that can go both ways: on the part of the narrator, it serves to observe the other – at a distance – as well as to evaluate them;[46] from the perspective of the other, it is a means to question and cross that distance and, moreover, force the traveler to 'look at' and reflect on herself, making her aware not just of her own body but of her own body in relation to the other and the other's space.[47] Traditionally a sense implying objectivity, and hence also a 'tool' in the creation of difference and hierarchies in much of travel literature, looking can separate and connect as these examples illustrate. When the tables are turned and the traveler is looked at, the comfortable, controlling position that the sense of vision often implies can quickly disappear, and the observing 'other' can move uncomfortably close.

A less hostile but all the more curious looking back can be found in Magoffin's account of the Santa Fe trail. When the narrator accompanies her husband into one of the villages, in fact, to see the Mexicans living there, her plans are somewhat thwarted since she herself turns into the object of attraction, a *"monkey show"* with which her husband could have made quite some money (92, emphasis in text). What follows is an interesting scene of mutual stares:

> My veil was ingenuously drawn down, not only for the better protection of my face from the wind and constant stare of *'the natives,'* but also afforded me a screen from whence to beholding my schrutinizing [sic] spectators, and while I carried on a conversation with Mr. Houk on the outside respecting them. There were some two or three dozen of children (both sexes) [...]; none were wholly clad, and some of the little ones in a perfect state of nudity; eyes were opened to their fullest extent, mouths gaped, tongues clattered, and I could only bite my lips and almost swallow my tongue to restrain my laughter.
>
> (92, emphasis in text)

The spectacle of the other works both ways here as the observer looks at the natives only to become the object of attention herself. Not only does she fail to hide her own gaze, but she also becomes a curiosity, awakening an unprecedented interest (open eyes, gaping mouths) in the locals that remind her of the 'monkey' or 'freak' shows that furnished Americans with unusual sights of 'the other' at that time. Returning and challenging the traveler's gaze, the travelees turn the tables and clearly define whose body is the 'other' in this scene and thus becomes a circus-like spectacle.

Another form of the traveler turning into a spectacle for the travelee can be found in a scene in Farnham's account, in which the 'indecency' and offensiveness of the lower-class settlers intersect with their overstepping of boundaries. In the "unequivocally filthy" room (Farnham 73), the ashes of the fireplace are blown around and also onto the table on which the house-wife kneads dough so that "the principal object of the woman's labor seems to be to distribute this brown coating [the dirt on the table] fairly through the mass [of dough]" (73). Mesmerized by the kneading, the narrator's "eyes grow to those balls of dough, and will not be persuaded from them" (73); her sense of vision – usually a means of distancing oneself from the scene – works differently here and draws her uncomfortably near to both people and food. The collapsing sense of distance is expressed in the disgust Farnham feels as the dirt renders the food inedible in her eyes (75), a poten-tial source of contamination that is expressive of all that is 'wrong' in this home. Again, visual distance collapses into an uncomfortable, unwanted proximity as disgust at the sight of the food – something potentially incor-porated and thus contaminating – links traveler and travelee.

The uncleanliness of the room finds its equivalent in the filthy and inap-propriate clothes of the family. While the husband "has on neither coat nor waistcoat, the whole of his simple garb being made up of a pair of blue cot-ton pantaloons, and a muslin shirt of very doubtful hue", his wife is "clad in a dark blue calico, made very short in the waist and very narrow in the skirt" (73). Like her husband, she covers her body only immodestly, which makes what she does next more convenient but no less shocking to the narrator: she breastfeeds her baby in front of everyone (74).[48] The whole family is ill-clad, and the dirty and unmatched clothes become an outward expression of their inward degeneration. Other details in the scene strengthen the impression: the degeneration goes more than skin-deep as in addition to unkempt hair, a lymphatic face (Farnham 73) or a very unfeminine "broad flat foot" (74) suggest that the lack of civilization is deeply and irrevocably 'embodied'.

This contrasts strikingly with the narrator, who not only has to suffer the family's sight but also their gaze and, moreover, their touch. The two daughters

> have politely turned their attention from ornamenting the loaves [of bread] on the table with finger marks, to the inspection of my dress and person. Let the reader figure to himself, then, the dirty house, the dirtier man, the dirtiest woman, and the most dirtiest children, for nothing but a double superlative will convey any idea of their condi-tion, and the writer sitting in the midst clad all in white of the most unsullied purity, [...] The whiteness of my dress seemed to amaze them. They took hold of it in various places, and lifted it from the floor to get a look at my feet.
>
> (74)

The family's dirty appearance stands in stark symbolic contrast to the nar-rator's white dress, which emphasizes the potential contamination that has become visible earlier. Apparently, the children have never seen (or touched) such whiteness, or so the narrator has us believe, which emphasizes this contrast, and the "unsullied purity" signified by the color appears all the more vulnerable.[49] And indeed, clearly violating the boundaries of respectability, the children touch the narrator and her dress, look at her feet and socks, and even handle her gloves (Farnham 75), which are supposed to protect her from just such a direct contact. The contamination threatened by the disgusting food becomes, in the words of the narrator, reality in the transgressive touch. Not only are the bodies here unregulated by the narrator's (middle-class) standards of decency, cleanliness, and hygiene, but, what is more, the 'traffic' between them (in the form of looking and touching) seems out of control as Farnham does not just turn from the observer into the observed but is also turned from a distant 'sight' into a close 'object' to be handled.[50]

Moreover, in a slightly different reading of the interaction, governed as it is by sight as well as by touch, one could say that the children resist the narrator's gaze and refuse to be turned into passive objects of evaluation (and amusement). In this as well as the other passages, look and touch on the part of the settlers cross the distance that the narrators try to create through their own scrutinizing and assessing gaze. Both the returned gaze and the touch serve to make the travelers aware of their own embodiment, turning them from a detached traveling 'eye' into a traveling body that is very much part of the frontier environment. It comes as no surprise that this passage ends with a description that turns the children into animals playing "*on all fours*" (Farnham 75, emphasis in text), a description that helps to increase the distance again, regulating the uncomfortable proximity between narrator and settlers. Additionally, the way this encounter between narrator and other is rendered narratively seems to stage the scene for us so that we – as readers – seem to become 'eyewitnesses' to this scene, looking ourselves – somewhat 'democratically' and without distinction – at every body, not just the lower-class one.

I would like to close with another scene that draws attention to such (bodily) practices of resistance with which the traveler's own intrusive gaze is met, coming from Kirkland's frontier account. When she accompanies her husband on an outing to inspect the future town that they want to found, she visits one of the settlers in the hope of something to eat. Greeted "with a civil nod by the tall mistress of the mansion" (Kirkland 13), the narrator finds the typical living arrangement, a one-room log-cabin which is sparsely furnished in a makeshift style. However, to her, it is a room that is in disarray, in which beds, wardrobe, kitchen utensils, and an oven share the space and which is, furthermore, tastelessly decorated with broadside sheets advertising a circus (13). The disorder of the room is matched, if not made worse, by the inhabitants' inappropriate clothing and behavior.

Apparently, the narrator has interrupted preparations for a meal, and upon welcoming her, the mother disappears promptly to change into a different dress (to "slick up" as she says (13)). The daughter, meanwhile, is kneading dough, wearing curlpapers in her hair (13). Slightly later, both women comb their hair right where dinner is served, in the words of the narrator "scatter[ing] the loose hairs on the floor with a coolness that made me shudder when I thought of my dinner" (14). To Kirkland, this mixing of categories reveals a body unregulated by principles of domestic propriety. As the women's private toilet mingles with their preparation of food, the increasing separation of body functions from other areas of life, which is a hallmark of the process of the disciplining regulation of human bodies,[51] is disregarded. Although the spatial (and, possibly, economic) restrictions in a frontier cabin do not allow for the neat separation of bedroom and kitchen, let alone afford the space for a separate parlor to entertain guests, Kirkland reads this spatial conflation as a violation of the standards of refinement and gentility, using it to mark the two settlers as 'other', different from herself.[52]

Nevertheless, when both women dress up for their visitor, apparently trying to diminish the class-based gap between them, Kirkland is very much surprised: "the young lady vanished – reappeared in a scarlet circassian dress, and more combs in her hair than would dress a belle for the court of St. James; and forthwith both mother and daughter proceeded to set the table for dinner" (14). Clearly, the narrator (who tellingly does not comment on her own clothes here) ridicules the women's attempt to dress up for her, pointing out the inappropriateness particularly of the daughter's overly fancy dress and hairstyle for a log-cabin in which the kitchen doubles up as a parlor and a bedroom. Moreover, the comment on the Circassian dress can be read as sarcastic: Circassian women from the North Caucasus, admired for both their gentility and their beauty, were associated with the 'Oriental' harem.[53] Connecting the young woman to the exoticized and eroticized stereotype of the 'Circassian beauty' does not just highlight her falling short of that ideal (she is just wearing the dress, after all); the connection to the highly sexualized context of the harem subtly suggests an inappropriate sexual behavior on her part, something that was often done with regard to lower-class women. In this sense, the woman's dress is doubly out-of-place, a reference to the exotic elegance of harem women in the middle of the lower-class milieu of the frontier cabin. Again, class difference is made to appear as embodied: while the fancy dress cannot cover the women's lack of refinement and education, it does reveal their inappropriate sexual character.[54]

However, while I see this passage as Kirkland's attempt to amuse her readers at the expense of the settlers, who appear uneducated and unrefined in the way they dress, behave, and more generally live, I would suggest that the scene can just as clearly be read as a form of bodily resistance on the part of the settlers. The women respond to their visitor's embodied class status, made visible by her dress just as much as by the fact that her husband owns the land on which the future town is planned. They dress up in an attempt to question and negate this status and the hierarchy it implies.[55] While this could also be considered as a form of positive acknowledgment – the women honor their guest by wearing their Sunday best – Kirkland certainly does not see it this way. She perceives it to be in bad taste, as revealing the fact that the gap between herself and these women is indeed unbridgeable; the women's gesture of empowerment, matching the guest's (more) elegant dress, is reframed by the narrator as a ridiculous attempt to appear elegant and well-dressed, and the lower-class body becomes once more guilty of transgression as dress, place, and person violently clash in the eyes of the narrator, turning into another spectacle of the unrefined lower-class body.

7.5 Conclusion

Particularly in the travel texts discussed here, an interesting paradox emerges between the freedom afforded by the frontier (welcomed by the writers for themselves) and the resulting lack of restrictions that can lead to a de-regulation of the bodies who live there. Thus, in addition to the usual sights and sceneries that have become a conventional part of travel accounts, the travel writers present us with what I have referred to as spectacles of otherness. Such views of deviant bodies serve to visualize the threat of decivilization that the travel writers so keenly feel, not least because the frontier space is not yet ordered (and thus stratified). The sense of seeing emerges as a privileged tool for encountering the other, supplemented by other sense perceptions such as smell or touch. As in other travel texts, sight functions as a conventional indication of truth, as evidence ("eyewitness") that cannot easily be refuted. Thus, vision links up with discourses of science and empiricism that were important at the time and serves to authorize female writers' voices and what they had to say. As I argue above, the sense of vision also functions as a distancing tool, widening the gap between observer (subject) and observed (object), constructing a sense of superiority that might not have been felt at the actual time of encounter (imagine Farnham being cornered by a group of children who keep touching her and her dress). At the same time, the sense of hierarchy created by looking (down) at the other is undercut in various ways. First of all, the observed takes the freedom to unabashedly look back, to 'answer' visually, thus minimizing the safe distance created by the evaluative gaze by reflecting it. Moreover, other sense impressions recorded as having been felt at the time of encounter similarly bridge that distance: the other bodies do not just return the look but also literally touch the travelers, hence moving into close proximity. Similarly, smell and noise can bring the other uncomfortably near, when body-related practices such as eating, drinking, or spitting are described. In this way, any absolute difference constructed by the visual spectacles presented in the text threatens to collapse, and travelers and travelees become ambiguously and disturbingly related to each other. The lack of regulations and the resulting disorderliness that can be seen with regard to the lower-class bodies encountered here become very much part of the depiction of the frontier, contrasting with the typical frontier narratives by male authors, whose emphasis on masculine heroics and adventure almost always ignore the role of the domestic sphere. Thus strengthening the ideology of domestication, these writers participate in the national discourse on the frontier and, with this, also elevate their own position for the nation's future.

Notes

1 Margaret Topping, "Seeing", in Alasdair Pettinger and Tim Youngs, eds., *The Routledge Research Companion to Travel Writing* (Abingdon: Routledge, 2020), pp. 193–207 (at p. 193).
2 Susan Shelby Magoffin. *Down the Santa Fe Trail and into Mexico. The Diary of Susan Shelby Magoffin, 1846-1847*, ed. Stella M. Drumm (Lincoln: University of Nebraska Press, 1962), p. 10. Further references to this edition are included parenthetically in the text.
3 Coming from a wealthy Kentucky family and married to a successful trader, Magoffin was able to travel in comparative comfort. Still, although a busy trading route, the Santa Fe Trail was part of the frontier and led through largely unsettled territory. What is more, the Magoffins traveled during the time of the Mexican War, often in the wake of the American army moving south, which made the journey even more risky. Still, judging from her diary entries, she seems to have thoroughly enjoyed traveling in that region.
4 Eliza W. Farnham, *Life in Prairie Land* [1846], ed. John Hallwas (Champaign: University of Illinois Press, 2003), p. 55. Further references to this edition are included parenthetically in the text.
5 Particularly from our perspective, the distinction between artificial and natural is problematic with regard to bodies and bodily practices. Nevertheless, it can be frequently found in nineteenth-century self-help literature. While the aim of such texts was, among other things, education and refinement of the 'civilized' body), their authors warned against too much 'artifice' in this area. The 'fine' line drawn here between refinement and artifice seems to coincide – at least for the thoroughly middle-class authors – with the distinction between middle- and upper-class values. Indeed, 'artificial' bodies can also be found on the frontier, where they appear as completely out of place. Kirkland comments on several characters who seem completely displaced, made unfit for frontier life because of the artifice coming from refinement and education. Trying to emulate an upper-class lifestyle on the frontier, for instance, is what renders characters unfit (e.g., Mr. B– and his family in chapt. XIX or Eloise Fiedler in chapt. XXVII).

6 Isabella L. Bird, *A Lady's Life in the Rocky Mountains*, intr. Daniel J. Boorstin (Norman: University of Oklahoma Press, 1960), p. 10. Further references to this edition are included parenthetically in the text.

7 Sarah Royce, *A Frontier Lady. Recollections of the Gold Rush and Early California*, ed. Ralph Henry Gabriel (Lincoln: University of Nebraska Press, 1960), p. 44.

8 Caroline Kirkland, *A New Home, Who'll Follow? Or Glimpses of Western Life* [1839], ed. Sandra A. Zagarell (New Brunswick: Rutgers University Press, 1990), p. 3. Further references to this edition are included parenthetically in the text.

9 Both topics would certainly be worth exploring but go beyond the scope of the chapter. While particularly Kirkland portrays a number of deviant male bodies, making frontier violence against women an explicit topic of her account (7), racially constructed otherness plays a lesser role in the texts discussed here. Native Americans are, in many cases, conspicuously absent; when they appear they are either silent, hardly articulate figures, or drunken and loud (29, 81, 85).

10 Mary Louise Pratt, *Imperial Eyes: Travel Writing and Transculturation* (Abingdon: Routledge, 1992).

11 The term 'travelee' is taken from Mary Louise Pratt, who defines it in analogy to the 'addressee' to highlight the importance of the perspective of those people encountered (and described) by the traveler: "'travelee' means persons traveled to (or on) by a traveler, receptors of travel" (*Imperial Eyes*, p. 242, fn 42).

12 For a history of the West that takes the idea of conquest (with all its economic implications) as a focal point in an attempt to criticize the 'myth' of the frontier and its ethnocentric and nationalist implications, see Patricia Nelson Limerick, *The Legacy of Conquest: The Unbroken Past of the American West* (New York: Norton, 2006).

13 Frederick Jackson Turner, "The Significance of the Frontier in American History", *The Frontier in American History* [1893] (New York: Henry Holt and Company, 1920), pp. 1–38 (at p. 3).

14 Richard Slotkin, *Regeneration through Violence: The Mythology of the American Frontier, 1600–1860* (Middletown: Wesleyan University Press, 1973); Henry Nash Smith, *Virgin Land: The American West as Symbol and Myth* (Cambridge, MA: Harvard University Press, 1970); see also Annette Kolodny, *The Lay of the Land: Metaphor as Experience and History in American Life and Letters* (Chapel Hill, NC: The University of North Carolina Press, 1975).

15 Annette Kolodny, *The Land Before Her: Fantasy and Experience of the American Frontiers, 1630–1860* (Chapel Hill: The University of North Carolina Press, 1984).

16 Brigitte Georgi-Findlay, *The Frontiers of Women's Writing: Women's Narratives and the Rhetoric of Westward Expansion* (Tucson: The University of Arizona Press, 1996), p. 12.

17 This ideology circulated notwithstanding the fact that the so-called separate spheres might not have been as separate and distinct as the wording suggests; writing, for one thing, worked as an expansion of female influence into the public sphere. For a discussion of the concept as part of the social fabric of the young nation, see Nancy Cott, *The Bonds of Womanhood: "Women's Sphere" in New England, 1870–1835* (London: Yale University Press, 1977); for its critical discussion see, for example, Linda Kerber, "Separate Spheres, Female Worlds, Woman's Place: The Rhetoric of Women's History", in Cathy N. Davidson et al., eds., *No More Separate Spheres! A Next Wave American Studies Reader* (Durham: Duke University Press, 2002), pp. 29–65. For a discussion of

true womanhood see Barbara Welter, "The Cult of True Womanhood: 1820–1860", *American Quarterly* 18, no. 2, Part 1 (1966): 151–74; for the concept of the Republican Mother see Ruth H. Bloch, "American Feminine Ideals in Transition: The Rise of the Moral Mother, 1785–1815", *Feminist Studies*, 4, no. 2 (1978): 101–26; Linda K. Kerber, *Women of the Republic: Intellect and Ideology in Revolutionary America* (Chapel Hill: The University of North Carolina Press, 1980); and Mary Beth Norton, *Founding Mothers and Fathers: Gendered Power and the Forming of American Society* (New York: Vintage, 1997).

18 Alva Noë, *Action in Perception* (Cambridge, MA: MIT Press, 2006), p. 7.

19 Noë, *Action*, p. 10.

20 In his "enactive" approach to perception, Noë conceptualizes perception as an activity: it "is not something that happens to us, or in us" but rather "something we do" (*Action*, p. 1).

21 Lisa Heldke, *Exotic Appetites: Ruminations of a Food Adventurer* (Abingdon: Routledge, 2003), p. 5.

22 This is also the case for seeing as John Berger points out: "The way we see things is affected by what we know or what we believe" (*Ways of Seeing* (London: BBC and Penguin Books, 1972) p. 8).

23 The fact that texts do not provide us with 'direct' access to anyone's experience (or the memory thereof) but are mediations and transformations that relate in complex ways to the genre of travel writing and the conventions and expectations surrounding traveling itself means, of course, that I cannot discuss the travelers' perceptions as such (at the moment these are made). Rather, I focus on how experience and memories are framed and transformed into texts.

24 Norbert Elias, *The Civilizing Process. Sociogenetic and Psychogenetic Investigations*, rev. ed., trans. Edmund Jephcott, ed. Eric Dunning, Johan Goudsblom, and Stephen Mennell (Hoboken: Blackwell Publishing, 2000), p. 52. See also Chris Shilling, *The Body & Social Theory*, 3rd ed. (Thousand Oaks: Sage, 2012), p. 162.

25 Shilling, *The Body*, p. 163.

26 Elias, *The Civilizing Process*, p. 471.

27 Shilling, *The Body*, p. 163.

28 Elias, *The Civilizing Process*, part 3.

29 Shilling, *The Body*, p. 167, emphasis in text.

30 Personal hygiene, e.g., became more privatized, something that relied, among other things, on the development of the toilet as a separate unit (Shilling, *The Body*, p. 167).

31 I would like to emphasize that 'civilization' in this context refers to a distinct set of (changing) Western and class-based body practices; despite these changes, these practices are often taken to be the 'natural' standard against which 'human' behavior is measured and evaluated. Civilization is, therefore, a very loaded term, as it has been typically used to refer to Western cultures and their ways of life in contrast to 'other' cultures that are then judged to be 'primitive'.

32 Berger, *Ways*, p. 9.

33 Laura Mulvey, "Visual Pleasure and Narrative Cinema", *Screen*, 16, no. 3 (1975): 6–18 (at p. 11). See also Berger, *Ways*, chapt. 3.

34 Pratt, *Imperial Eyes*, p. 7.

35 Pratt, *Imperial Eyes*, p. 7.

36 Giorgia Alù and Sarah Patricia Hill, "The Travelling Eye: Reading the Visual in Travel Narratives", *Studies in Travel Writing*, 22, no. 1 (2018): 1–15 (at p. 1).

37 "Grenzüberschreitende Blicke: Wahrnehmungen 'bewegender' Körper", in Tanja Gnosa and Kerstin Kallas, eds., *Grenzgänge. Digitale Festschrift für Wolf-Andreas Liebert* (2009), pp. 1–18 (at p. 14).

38 See, for example, Ansgar Nünning, "Zur mehrfachen Präfiguration/Prämediation der Wirklichkeitsdarstellung im Reisebericht: Grundzüge einer narratologischen Theorie, Typologie und Poetik der Reiseliteratur", in Marion Gymnich et al., eds., *Points of Arrival. Travels in Time, Space, and Self. Zielpunkte, Unterwegs in Zeit, Raum und Selbst* (Tübingen: Francke, 2008), pp. 11–32. There is a difference between the traveler and the travel writer, more precisely, between the traveler's actual experience at the time of travel and its later transformation into a text, a transformation that is mediated not just by the passage of time (between experience and writing) but also by aspects such as conventions of traveling and travel writing, purpose and expected audience of the text, etc. In the following, my focus lies on the travel writer and the ways in which she renders and frames her act of traveling.

39 See, for example, Clare Brant, "Smelling", in Pettinger and Youngs, eds., *Companion to Travel Writing*, pp. 249–61, for an overview of how smells and odors can serve to order the world, to mark differences and boundaries (esp. 253–55).

40 See also Meyer's contribution on Mary Wollstonecraft's travels in Scandinavia in this collection, in which he discusses a similar gesture of 'othering' the poor body by emphasizing the lack of hygiene.

41 Paul Rozin and April E. Fallon, "A Perspective on Disgust", *Psychological Review*, 94, no. 1 (1987): 23–41 (at p. 24). Rozin and Fallon discuss disgust as a "food-related emotion" (23) which is elicited by the anticipation of having to ingest "an offensive" object that is seen as contaminating or contaminated (23). Although here, the other bodies are not literally meant to be eaten, one could argue that their bodies are seen as 'contaminated/ing' and thus, offensive. Moreover, they are in so far 'incorporated' as they are (already) part of the nation, and hence, something that is potentially dangerous to the national well-being. See also Schaffers in this collection for a discussion of disgust in relation to food and incorporation during travel.

42 Moodie's narrative deals with the Canadian frontier and was therefore not directly part of the US-American discourse; yet similar images and ideologies were at work here, which justifies the inclusion of her narrative.

43 Susanna Moodie, *Roughing It in the Bush* [1852], ed. Michael A. Peterman (New York: Norton, 2007). Further references to this edition are included parenthetically in the text.

44 Tellingly, in an earlier passage, Moodie describes the neighbor's daughters as "[f]ine strapping girls […] but rude and unnurtured as so many bears" (91), which fits to her comment in this scene that they seem to sleep together with their dogs.

45 Sarah Jackson, "Touching", in Pettinger and Youngs, eds., *Companion to Travel Writing*, pp. 222–35 (at p. 223).

46 Alù and Hill, "The Travelling Eye", p. 1.

47 Pratt has already pointed out this mutual regulation through the (returned) gaze: "the two sides determine each other's actions and desires" (*Imperial Eyes*, p. 80). See also Alù and Hill, who write that vision and writing "are mediating and embodied actions through which travelers have made and remade the world around them" ("The Travelling Eye", p. 3).

48 While breastfeeding attained, at the time, an almost 'sacred' character, as it was thought to embody an ideal form of motherhood, it came with complicated

race and class associations. These, together with the visible and intimate nakedness displayed in the scene, might be the reason for the sense of shock Farnham feels here. See, for example, John C. Waller, *Health and Wellness in 19th-Century America* (Santa Barbara: Greenwood, 2014), esp. pp. 102–104; Cornelia C. Lambert, "Breastfeeding", in Marilyn J. Coleman and Lawrence H. Ganong, eds., *The Social History of the American Family: An Encyclopedia* (Thousand Oaks: Sage, 2014), pp. 144–47; and Janet Golden, *A Social History of Wetnursing in America: From Breast to Bottle* (Columbus: Ohio State University Press, 2001), esp. chapt. 2 and 3.

49 Similar to the scene in Moodie's travel narrative discussed earlier, the question of who does the washing is pertinent here. While the fronterswoman under scrutiny here certainly has to do her own washing, Farnham in all likelihood had her clothes washed by another woman and thus did not have to worry about the state of her clothing. The travel writers discussed here might not be rich, yet it is likely that they do not do their own washing, as some of their comments suggest (Kirkland, for instance, makes finding appropriate domestic help a topic in her account; 38–40, 42, 51). Washing was one of the most strenuous duties in the household so that the idea of wearing white for women who had to clean their own clothes might indeed be unthinkable. Interestingly enough, Moodie, who in vain tries to find domestic help, describes the gruesome – and difficult – task of washing her own clothes until her hands are bloody. Nevertheless, she, too, takes clean clothes for granted on the frontier and makes them a measure of respectability.

50 As in the passage recounted by Moodie above, this scene hides that the narrator here is also an intruder into the settlers' private sphere. During the construction of their own houses, Farnham and Kirkland are staying with a variety of settlers for their own comfort so that, in a certain sense, their evaluative gaze (and certainly their public comments on their hosts) is a transgression of hospitality.

51 Elias, *The Civilizing Process*, pp. 114–21.

52 However, Kirkland also makes fun of such preconceptions when she describes her own effort to set up house, trying unsuccessfully to fit parlor furniture, elegant china, and many other of her beloved but superfluous things into her own log-house under the incredulous stares of her neighbors (41–44).

53 Joan DelPlato points out that "[b]y far the most frequently mentioned ethnicities or 'races' of harem women were Circassian or Georgian associated with the Caucasus mountains" (*Multiple Wives, Multiple Pleasures: Representing the Harem, 1800–1875* (Rosemont: Rosemont Publishing and Printing Corp/ Associated University Press, 2002), p. 39). She also discusses a number of nineteenth-century paintings and texts that referenced Circassian women and their beauty in the context of the harem.

54 Indeed, Kirkland hints at the 'sexual deviance' of some of the settlers with one of the daughters even dying because of her pregnancy (chapt. 13).

55 Such a reading fits other passages in Kirkland's travel account, in which the narrator's class-based sense of difference and superiority is undermined by the settlers' sense of democratic equality; they frequently refuse to behave in deferential ways and thus to acknowledge the superiority that Kirkland feels.

III

Crossing Borders

The Body and Its Liminal Zones

8 "My Condition Gets Worse Day by Day"[1]

Controlling Traveling Bodies on the Move in Edo-Period Japan

Andreas Niehaus

8.1 Introduction

Nomi shirami	Fleas and lice
Uma no bari suru	A pissing horse
Makura moto	Next to my pillow[2]

The famous poet Matsuo Bashō (1644–94) penned down these verses in his *Oku no hosomichi* (*Unbeaten Tracks in the East*), a poetic travel diary in which he gave an intimate and rare glimpse into the corporeal realities of his journey through Northern Japan in 1689. Bashō traveled, as was most common, by shanks' mare, but due to rain and heavy winds, which rendered traveling impossible, he was forced to stay in the private house of a checkpoint official for a period of three days: "against our will we had to stay in this inhospitable mountain region".[3] The situation Bashō is forced into places him outside of his emotional and bodily comfort zone (proximity to animals, the sound and the stench of the horse's urine[4]) and, more significantly, also threatens his health due to unhygienic circumstances (bites of fleas and lice). Bashō's diary draws the reader's attention to the fact that traveling can challenge physical awareness, or, put differently, will make the traveler aware of his/her body, which in daily life tends to disappear under layers of bodily routines. Being placed in the out-of-the-ordinary situation of traveling certainly offers opportunities to free oneself from social (gender) and political constraints and to invent oneself anew. However, being away from home and one's daily routines also means to be confronted with and being forced into unfamiliar and unwanted situations, which bear the potential of a loss of control over one's body. In this article, I will analyze exemplary passages in Edo-period travel writings that deal with bodies losing, trying to maintain, and regaining control. As traveling happens at the intersection of historical, geographical (spatial), cultural, social, economic, and political conditions and circumstances, which will influence the experiences of the traveler (and the comprehension of the reader), I will briefly introduce the historic-cultural context of traveling in the Edo-period.

DOI: 10.4324/9781003331803-11

8.2   Travel Writing in the Edo-Period: Towards Modernity on Shank's Mare

The Edo-period (1603–1868) witnessed a travel boom from the turn of the eighteenth century onwards, which resulted in a broad array of travel texts.[5] As diverse as the reasons for travel (like forced or voluntarily, official, scholarly, poetic, recreational, or as a pilgrimage) were the reasons to write, and therefore the literary characteristics of travel texts stretch from mainly factual botanical or geographical documentations, records of customs, local legends or myths, travel guides, and road maps written for practical use to different degrees of fictional representations in prose, (poetic) diaries, and novels.[6] Recently, it has been argued that the turn towards a "more realistic type of travel writing"[7] was triggered by an increased interest of Neo-Confucianists in Chinese pharmacology and medicine (*kampō*). Within Neo-Confucianism, the "study of nature was considered a socially beneficial, pragmatic endeavor, and a sufficient means to cultivating the self. Since all things *(ki)* were understood as participating in the ultimate truth *(ri)*, study of the natural sciences was thought to serve as an intermediary in man's search for the truth".[8] The "search for the truth" meant that intellectuals left their studies. They traveled across the country to observe, measure, and describe landscapes, history, plants, animals, clothes, customs, legends, etc. In this respect, the travel diaries of these intellectuals represent a transition towards modernity; a transition that is accompanied by a new way to observe and by an attempt to expand the boundaries of one's knowledge: the authors approached their environment no longer with an uncritical attitude and the intention to (intellectually or physically) mainly revisit famous places (*meisho*) for confirmation and the sake of tradition. In his travel diary *Saiyū zakki* (*Notes on a Trip to the West*, 1783), Furukawa Koshōken (1726–1807), a physician and geographer, complained that the travel guide he carried did not give the exact distances between villages.[9] Travelers were thus not only traveling with certain ideas about the places they wanted to visit, but they brought earlier travel texts along, comparing text and visual, as well as emotional impressions in situ. Intellectuals like Koshōken traveled with the critical eye of the scientist and explorer and with the intention to check the reality they experienced with their bodies – with what they saw, smelled, heard, and touched – against old texts and handed-down knowledge. Moreover, they came equipped with the necessary instruments that enabled them to measure and compare, and in the end to create a new reality. Western instruments like telescope and microscope "profoundly altered Japanese constructions of the meaning of sight".[10] This new gaze was also what the reader expected so that especially the increasing number of travel guides had to be exact descriptions of the reality a traveler could and would expect to find. Records in the tradition of the Confucian scholar, botanist, physician, and enthusiastic traveler Kaibara

Ekiken (1630–1714) were all about the realistic representation of places visited, customs observed, food eaten, and legends heard. For the audience, the presence of the traveler's or writer's body and the lived experience then guaranteed authenticity vis-à-vis the written accounts of the past.[11]

However, we also find a broad array of fictitious travel writing as well as poetic travel diaries. For poets like Bashō, who became an Edo-trendsetter in poetic travel diaries, (poetic) journeys were still meant to bring the poet into contact with the past; and with his journey to the East he wanted to follow the footsteps of the *waka* poet Saigyō (1118–90).[12] However, in contrast to earlier travel writings, which often meant 'dropping' *meisho* (famous places) and *utamakura* (poetic pillow words) in the text without physical journeying and thus without actually leaving the safe environment of one's study, Bashō, as an Edo-period travel writer, had to experience the places in reality by using all his senses.

Edo-period travel writers wanted to express and represent the reality of the visited places as well as the immediacy and emotions experienced during travel. This is important as the body becomes the "site of translation between the material and inner worlds, the environment and the self", and the senses then "indicate and shape how one travels".[13] Bodies and bodily sensations and experiences are no longer merely metaphorical markers that do not refer to any real experience, but rather evolve into a text element that wants to transmit an experienced (somatic) reality. This new approach also asks for a new language, which is characterized by the choice of concise words and colloquial Japanese instead of a formal and poetic language.[14] The transmitted realities are then discursive expressions of a "pre-reflective correspondence between body and world"[15] and as such, they merely mediate corporeality. Yet, there is also a reciprocity between the experience of the body and the text, opening a space to analyze how corporeal experience is expressed and given form in travel writing.[16] While several studies have been published on traveling, travel literature and traveling and representation in Japan,[17] studies on Edo-period traveling bodies and on the reciprocity between the experience of the traveling body, experience, and text are so far missing.

8.3 The Historical and Cultural Context of Traveling in the Edo-Period

The society of the Edo-period was feudal and based on a strict separation between the four classes: warriors, peasants, artisans, and merchants. The system of balancing power between the ruling Tokugawa family and the local lords (*daimyō*) included a hostage system, which stipulated that, for example, female family members of lords had to stay as hostages in Edo (today's Tōkyō), the center of political power. Additionally, local rulers had to travel to Edo in regular intervals as part of their military duty (*sankin kōtai*).[18] While official traveling was thus an act that contributed to the stability of the political order, the central government at the same time identified 'uncontrolled movement' of the population as a threat and as a destabilizing factor that opened the door to political conspiracies and social uprisings and therefore aimed to control access to and movements on the main roads. Very early in the Edo-period an infrastructure, whose maintenance was institutionalized by the authorities, was established that had to serve as a mechanism to order the 'chaos', to control movements, and to create official symbolic spaces. The government heavily invested in the five main roads (*gokaidō*), among which the Tōkaidō (Eastern Sea Road) – connecting Edo and Kyōto – was the main and most frequented road. Being a rather small and not well-maintained road before the Edo-period, it was developed into a road with a width of nine meters, with little hills as distance markers (every 3.927 km) and trees planted to provide shelter during the summer and protection from snow and wind in the winter. Official inns for *daimyō* (feudal lords), called *honjin*, were built to serve the needs of the lords that were forced to be present in Edo as part of their military duty.[19] Additionally, rest or service stations (*shukueki*) for official travels were constructed in an attempt to allocate space according to social hierarchy. However, these service stations soon expanded and served the needs of other travelers as well. Additionally, a total of 53 control posts/checkpoints (*sekisho*) were built by the government around Edo and between the provinces in order to control who was traveling why, whereto, and when. Travelers were meticulously checked and had to present a travel permit or a passport to cross to the next province.[20]

Roads were not just socially controlled, but also gendered spaces with gender being a major parameter to define the position of an individual in Edo society, and female journeying in the early Edo-period was mainly enforced travel in the context of political control mechanisms. However, starting around the turn of the eighteenth century, with the advent of a specific urban culture and with prosperity and stability came the invention of leisure so that men and women from different social background started to travel for a variety of reasons: pilgrimage, sightseeing, education, convalescence, etc.[21] The legal and official frame had defined women as passive and functionalized in the context of a male-dominated society. However, hitting the road for women could not only mean to break free from their daily life routine (as it would also for men) and the boundaries of their social class, but it also signified an escape from the corset of their gender roles and their limited opportunities to discover and experience the world. Commoners, men and women alike, increasingly went on pilgrimages in order to 'escape' their social restrictions, and in the case of peasants, being on the road meant an escape from their forced immobility (*nuke-mairi*, escape pilgrimage). Contemporary records speak of several mass pilgrimages to Ise. In 1705 about 3.5 million pilgrims went to the Ise shrine between the 9th of the fourth month and the 28th of the fifth month alone.[22] Pilgrimage leveled the difference of class as the pilgrim left behind earthly belongings and aspirations, indicated already by the uniform and modest garments. The military government thus struggled to control and contain movements on the roads by creating official spaces that allowed only limited access according to gender and status – but space (and travelers) resisted: on roads, "structures broke down, exposing both danger and freedom"[23] and "malleable landscapes rendered the enforcement of official directives on the topic [the debate on mobility] discontinuous and heterogeneous at best".[24]

8.4   At the Checkpoint: "I Still Felt My Heart Racing Once Exposed to Such a Forbidding Atmosphere"

> The woman inspector did a body check on me, combing through my hair, as well. She was a coarse and uncultured woman, and though old, quite healthy and strong. She sat closely by my side and questioned me in a throaty voice. I was frightened and nervous, wondering what would happen next. [...] While I had been sure that there had been no error on the travel permit, I still felt my heart racing once exposed to such a forbidding atmosphere. When the inspection ended, therefore, I was extremely pleased and immediately had the attendants gather to leave the site together. I had my hair fixed, when the party reached the top of the hill.[25]

This vivid and rare first-hand account of the control at the Hakone checkpoint is recorded by Inoue Tsū-jo – the daughter of Confucian scholar Inoue Motokata – in her diary *Kika kikō* (*A Homeward Journey*), while traveling from Edo to Marugame in 1689. The checkpoint of Hakone, situated on the Tōkaidō road, the main access point to Edo, was notorious for being rather strict. This alone would explain why Tsū-jo experienced the situation as "forbidding" and why she felt nervous, even fearful. Tsū-jo as a female traveler from the warrior class experiences the controls as a moment of being powerless and being at the mercy of checkpoint officials. Tsū-jo's anxiety, her loss of control, results in a somatic reaction: "my heart racing". The barriers were also a space where the privileged ruling class was treated more harshly than the commoners, especially women. The given political frame (female hostages in Edo) meant that traveling women were controlled more strictly when leaving Edo, and a saying during the period describes the priorities of the checkpoints as 'inbound guns and outbound women' (*iri-deppō ni de-onna*).[26] Women's travel diaries, not only Tsū-jo's, regularly comment on the treatment by officials at the checkpoints. Tsū-jo's uneasiness and fear were also not entirely unfounded as laws were strict, and crossing a barrier without permit could, be punished by crucifixion or deprivation of legal domicile. Tsū-jo's diary shows that one of her main concerns was her travel permit, despite the fact that she "had been sure that there had been no error". Her worries concerning her official travel documents become even more understandable when reading her diary *Tōkai kikō* (*Travel along the Tōkaidō*, 1681), which documents her journey to Edo some years earlier. Tsū-jo at that time was summoned to Edo to teach the mother of the lord of the Marugame domain, but at the checkpoint in Arai, she and her party were not allowed to travel further. Travel permits were very detailed in the information they contained. Tsū-jo was identified as an adult woman (*onna*) in her permit. However, there was a discrepancy between the information provided in the permit and the clothes she was wearing. In Edo-period society, it was possible to identify the age and marital status of women by clothing, and as Tsū-jo was wearing a kimono with long sleeves, she was categorized as a young woman (*ko-onna*), which was a different category in the permits. As a result, the travel party had to stay in Arai for six days until a revised pass was issued and her party finally could continue their journey.[27]

Yet, there is another reason why Tsū-jo experiences the situation as "forbidding", and this is related to the unwanted direct contact with an unknown woman from the lower classes (the male official could not even address her, let alone conduct a body search). The inspector is by appearance described as "coarse" and "uncultured", and also her "throaty" voice characterizes her as a member of the lower classes. The contact with the female inspector is first of all scandalous as Tsū-jo is forced to answer the questions of a female person that has a much lower status. Yet the main scandal and source of anxiety is the closeness and the enforced body contact with the female inspector. In her diary, Tsū-jo mentions that the inspector is sitting "closely" by her side, which would already be a situation out of the ordinary. Members of the female upper class during the Edo-period lived in 'protected' and often secluded and controlled spaces in which direct body contact even with household members was limited to certain moments of their daily routine: dressing, female toilet, and marital sexual contacts. These contacts were regulated performances firmly placed within the hierarchies of class and gender. For Tsū-jo, then, being touched and searched by a woman who was in her eyes "uncultured" would breach the frame that defined her identity as a woman of the upper class. The ordered world of normality as well as Tsū-jo's physical and mental well-being are only restored when she takes over control again ("immediately had the attendants gather") and after she had left the checkpoint, a space over which she did not have control. Tsū-jo's mental state as well as the destruction of social order is metaphorically reflected in her untidy hair. For Tsū-jo, it seems necessary to especially stress that the body check included combing through her hair, which certainly signified a transgression in terms of intimacy and shame. The act of hairdressing ("I had my hair fixed"), which immediately follows their departure from the checkpoint, then, not only formally and symbolically reinstates her power within the given social order, but also signifies her return to her 'safe space' and the 'freedom' which she receives from that social order.[28]

On the road, clear hierarchies as well as the separation of class and gender would certainly become perforated, and several fictional texts – with Jippensha Ikku's *Tōkaidōchū Hizakurige* (*A Shank's Mare*) being the most famous one – play with the potential fluidity of gender and class hierarchies on the road. Particularly crossings were a moment in which women of the lower classes would be forced to get into close contact with the other gender: for crossing rivers, one could hire different forms of 'portable platforms', but the cheapest was piggy bagging on a porter's shoulders, which meant very close body contact with scarcely dressed men (see Figure 8.1). Tsuchiya Ayako's diary *Tabi no inochige* (*Journey with a Writing Brush*) describes her experiences at the Ōi river crossing during a journey she had to take in 1806. Her husband had been posted as Sakai magistrate, and as

Figure 8.1 Crossing at Okitsu, Okutsu river, in Hizakurige gajō, *Tōkaidō meisho*. *[Bishō dōjin; Tamenobu gahaku]* (Tōkyō: Takamizawa Mokuhansha, 18--(year unknown), p. 25, https://www.library.dartmouth.edu/digital/ digital-collections/dochu-hizakurige).

his wife, she had to move to the city of his new assignment. She traveled along the Tōkaidō and gives a personal account of the places visited and the encounters on her way. Especially vivid is her description of the Ōi river crossing, which shows how crossings function as an intersection of class and gender. Ayako describes the river as "an intimidating site"[29] and she was carried over the river – as her position required – in a palanquin by hired porters:

> On the vast shallows were scary, coarse looking men, half naked, gathering around me. They quickly roped my palanquin, held it above their shoulders, shouted 'Yo-ho!' and sped across the dry riverbed like flying birds. It felt as if I was half dreaming, with my head and feet swinging around, feeling dizzy and having difficulty breathing. [...] Presently they ran ashore, then rotated my palanquin and quickly lowered it. I felt as if I had been pushed back to fall into a ravine thousands of meters deep. When the river crossing had been completed, the women were all so pale that the people around us laughed loudly.[30]

Just like Tsū-jo, Ayako has to relinquish control, and both women's bodies reacted to the out-of-the-ordinary situation ("feeling dizzy", "difficulty breathing", "the women were all so pale"). As they were on official, that is ordered journeys, both women were forced into situations in which they felt helpless, and while Tsū-jo reacts to the loss of her social self-concept, Ayako reacts to a potentially dangerous moment – a moment, which was clearly not considered to be especially dangerous by the male porters or bystanders. Their laughter ridicules the psychosomatic stress reaction but also serves as 'revenge' for those that are without social and political power. However, it can be argued that the act of laughter also reproduces gender hierarchies and gender stereotypes. This is precisely what Yasumi Roan's advice in the *Ryokō Yōshinshū* (*Precautions for Travelers*, 1810), a manual of do's and don'ts during travel – seems to suggest:

> Women are, in contrast to men, much more sensible when staying at the banks of a river and get frightened because of the force of the waves or the bustling activities all around. It can then happen that they feel dizzy and have problems with their circulation. Therefore, one should already point out the night before that there will be a lot of commotion, but that one should not be impressed by that.[31]

8.5   Forced Situatedness: "Tears Roll as Big as Acorns"[32]

My back hurt and the cold morning breeze from the valley was blowing in. I felt wretched but decided to endure the pains of travel and felt it was still better riding on a palanquin than creeping up the path on foot.[33]

Only to a certain degree could the means of traveling be chosen freely during the Edo-period. Tsū-jo – as required by her status and the occasion of her journey – was traveling in a palanquin, and when she was called over by the official at the control post, she did not walk over, but "had" her litter "moved towards the guard".[34] Renting a horse,[35] or being carried in a sedan (*kago*) or palanquin (*norimono*) were means of traveling that had to be paid for or were even prohibited entirely for certain classes. While a palanquin or a sedan, to the modern reader, seems rather comfortable, Norinaga's "pains of travel" are an example of the downsides of riding in a palanquin (including 'seasickness'), and Western travelers to Edo regularly complained about the 'imposition' of having to ride in a litter in their travel accounts. Philipp Franz von Siebold considered the palanquin to be rather uncomfortable as Europeans were not used to the Japanese body technique of sitting with crossed legs: a problem that meant a 'real torture' especially for taller Europeans.[36] Renting a cheap sedan likewise had its dangers as Kita in the *Tōkaidō hizakurige* had to experience when he rented a sedan whose bottom came out, sending him to the ground.[37] Commoners could by law not use certain means of transportation and were often forced to walk due to financial reasons, whereas pilgrims and writers on a poetic journey generally preferred to travel on shanks' mare voluntarily as it placed their endeavor within the tradition of ascetic exercises (*shūgyō*).[38]

The different modes of 'having to be' on the road certainly changed the situatedness of the traveler, and especially accounts of travelers on foot give a vivid account of the close and tactile proximity to the surrounding space: the smells, the sounds, the soil, the plants, fellow humans, and animals. These texts, then, also bear witness to wanderers being exposed to situations that were beyond their control. One of such situations would be – as also seen in Bashō's poem quoted at the beginning – being exposed to excrements of men and beast alike. A lord being carried in a palanquin or a rider on horse could easily (try to) ignore the excrements along the road, whereas porters and other travelers on shank's mare could not. Steep slopes, especially off the beaten tracks, could also mean that wanderers had to use their hands for climbing and would therefore be even closer to the ground. At some places along the Tōkaidō, droppings must have been a common sight, thus adding also to the smellscape of the traveler. One of these places was the steep Hakone slope, for which Asai Ryōi in his influential *Tōkaidō meishoki* (*Travel Record of Famous Places along the Tōkaidō*, around 1660) remarks that it provides the fertilizer for Hata.[39] In a humorous poem (*kyōka*), he further verses:

Shi-ri nobori	Climbing up four miles
Tani yori tani ni	Walking
He megurite	From valley to valley
Ni-ri ni-ri kudaru	Descending two miles, two miles
Hakoneyama kana	That is Mt Hakone[40]

While the translation of this poem does not include any reference towards droppings at first sight, the second layer of interpretation hints at a rather 'dirty descent': *shi-ri* does not only mean four miles but also "ass", *he* is a homonym for "flatulence" and *ni-ri ni-ri* (two miles, two miles) is also the onomatopoetic representation of diarrhea.[41]

While the upper class was thus able to travel in 'relative' comfort, commoners and especially those without financial means were obliged to walk, which was the most tiring form of traveling, and Ryōi also comments on common travelers losing composure due to exhaustion:

Kashinoki no	Crossing the hills
Saka wo koyureba	with the oaks,
Kurushikute	is exhausting,
Donguri hodo no	and tears roll
Namida koboru	as big as acorns.[42]

Bad conditions of the road due to rain and wind or tired and aching feet could not only slow down the traveler but also force him, like Bashō, to stop and spend days of faineance and boredom at inns or private homes. Koshōken stranded after a bridge had been washed away by heavy rain, and Sugae Masumi (1754–1829) was forced by wind, rain, and high-water levels to seek shelter repeatedly and on one occasion could only cross a river in Kanegasawa by holding hands with two other travelers supporting each other.[43] Thus, being on the road brings the traveler into closer bodily proximity with other travelers, forming a temporary community of support within a sphere of uncertainty. Traveling in company, as Kita and Yaji (*Tōkaidōchū Hizakurige*), Rokuami and Mokuami in Asai Ryōi's *Tōkaidō meishoki*, or Bashō and Kawai Sora (1649–1710) in the *Oku no hosomichi* was common and even recommended in travel guides. Yet the proximity to other travelers was double-edged: while giving security, it also added to the dangers of the journey. Travelers would be tricked or robbed[44] by other travelers on the road, and it seems that especially inns at the roadside had a bad reputation, as travel advice concerning precautions one should take in an inn shows.[45] The potential of conflicts at inns, where rooms had to be shared and where travelers were lodging close to each other, is especially taken up in travel writings.[46]

As travel writings show, traveling on foot meant exhaustion, and, whereas at home one could rest and regain strength from one's daily work in one's own familiar surroundings, choosing one's accommodation during a journey was not always possible, and one depended on accommodations at hand or the help of locals that offered a corner in their house to sleep.[47] The exhaustion of traveling on foot for an entire day made night rest a crucial element in the traveler's well-being and health.[48] Yet, the private places and inns for travelers were not always spaces that provided the necessary environment for rest, as Bashō's poem points out. Especially roadside inns for commoners were places where travelers (due to loosened social control) would 'whoop it up', and where drinking, parties, and prostitution would make a good night's sleep impossible.[49] The lack of sleep and the difficulty to fall asleep due to weather conditions, 'exotic' food, insects, and uncomfortable bedding[50] as well as an occasionally disturbing or uncommon soundscape are repeatedly penned down. During his travels, Masumi regularly encountered sleepless nights as the environment eludes his control, and with some frustration, he comments on his nights and the different reasons he cannot find sleep: "I moved my bedding away from the leak, put on my straw raincoat, lit the torches, and drew myself near the dying fire. The wind was again blowing fiercely and shaking the house. I could hear the men who got up and shouted: 'Wind abate!'"[51] Whether Masumi found sleep that night we do not know, but what is apparent is his relief when the accommodation provided protection:

Ka no kuwade	How pleasant,
Raku wa ika hodo	inside the net
Kaya no uchi	mosquitos are not biting[52]

8.6 Controlling One's Body's Health on the Road: "My Condition Gets Worse Day by Day"

My condition gets worse day by day, and I can hardly carry out daily activities. Being over a thousand kilometers away from home, I grieve over my old body that may not be able to survive [...] I asked where the wind was blowing from. They said it was from the west. I felt deeply homesick and kept on looking at the moon until it declined.[53]

These notes can be found in the *Akikaze no ki* (*Autumn Winds Diary*), written by Kyōto poetess Shokyū-ni. Shokyū-ni followed Bashō's footsteps to the East of Japan and had to stay in bed for more than 40 days due to a severe illness. This passage in her diary reminds us first that the body is the medium of travel, but secondly that the body potentially sabotages traveling, constraining the traveler's ambition to reach his/her goal either temporarily or ultimately. The body in its vulnerability, its weakness, and its limitation(s) therefore constantly forces itself *as body* upon the attention of the traveler. Travel writing reflects the immediacy of the bodies and the need to keep control over one's body's health, to keep the body fit for traveling. The texts give an intimate insight into the bodies' geography, the way bodies are experienced on the road as they show how travelers inwardly 'listen' to the condition of their own bodies and how they quarrel with their bodies' ailments. Travelers constantly tried to create the perfect environment for their traveling bodies within circumstances that are far from being perfect: inns for commoners and private houses along the roads were dirty and infected with fleas and lice, food in certain periods and areas was scarce, robbers, wild animals like boars and wolves, flooding, heavy rain as well as cold and heat were additionally threatening one's physical inviolability.

Increasing travel activities and the need to be prepared for journeys made travel guides (very often in pocket size) a popular genre. These guides – together with health and household manuals – reacted to the real and imagined dangers of traveling and the worries and questions of the traveling layperson by giving detailed information and advice concerning treatments for aching feet, footwear, the crossing of rivers, on codes of conduct, what to do in case of illness, choosing travel companions, travel equipment, etc.[54] These texts thus give us an idea of how travelers understood their bodies and their bodies' functions and how they hoped to control the well-being of their bodies. While provincial barriers politically controlled the travelers' movements, travelers had to control what entered and left their bodies through the skin as well as through the mouth. This act of controlling the exchange, or the balance, between body and environment is also political, as ideas of the body mirror Neo-Confucian ideas of self-control and self-government. Neo-Confucianism became state ideology during the eighteenth century and its metaphysical speculations and ethics were translated into "pragmatic and functional guidelines"[55] that disseminated into daily routines and cultural practices of the general population. Neo-Confucian metaphysics and ethics propagated a healthy and ethical body within the context of a political and social order that in the end reflected cosmic order. Individual health care, self-control, and self-cultivation were defined as duties to one's parents (filial piety), the state, and the ruler, and as the responsibility for being healthy was defined as a responsibility of the individual, illness was as well. An individual's health condition in that respect thus reflected one's 'ethicalness'.[56]

According to the teachings of contemporary Neo-Confucianism, Taoism, and *kampō* (Chinese) medicine, travelers would be prone to illness due to the harsh physical activity of walking long distances, which would cause an imbalance of the *ki* flow within the body.[57] While moderate walking was seen as a recreational activity, walking long distances caused an imbalance in the *ki* flow within the body and would lead to sweating and the opening of the pores, which then would make way for the elements wind, heat and cold, rain, and snow to penetrate the body. Weather could thus not only force the traveler to interrupt the journey but was also identified to be a major cause of illness. In his *Yōjōkun (Nourishment of Life Rules*, 1713),[58] the Confucian scholar, physician, and enthusiastic traveler Ekiken calls the elements, which are wind, cold, heat and dampness, the outer evils *(gaija)*, which cause severe ailments and death.[59] The outer evils enter the body via the skin, which is seen as a liminal organ that interacts with and reacts to the outside.

While the pores were believed to be closed in winter, they supposedly opened up in spring. Especially dampness was identified as the cause of different deadly ailments as it penetrates deep into the body and is therefore difficult to heal. As Siebold observed during his travels the "otherwise hardened Japanese display such a sensitivity to dampness".[60] Ekiken therefore advises to protect oneself at damp places like rivers, and one should, for example, not sit on the ground close to rivers.[61] The *Zōho shūgyoku chie no umi (Addition to the Collected Pearls of the Sea of Wisdom*, 1747) thus very precisely instructs: "Dampness can easily invade the body through the navel. One should therefore place two or three peppercorns on the navel, wrap a (under)belt around and walk through the river. One will be free of dampness."[62] Also household manuals recommended a variety of measures against the outer evils in order to control the energy balance between self and environment. The advice given not only included provisions concerning clothing or which places to avoid, but also advice regarding food: "In the morning before departure one should take a piece of ginger or drink pepper-powder cooked in hot water. That guards against heat in summer and cold in the winter"[63] and "[b]efore departure early in the morning, one should place ginger in one's mouth to be guarded against all harmful vapors, mist, dew, dampness, and cold mountain wind and not to fall ill".[64] Food, as these quotes make clear, was considered to be medicine. Food in this approach was thought to have the quality to rebalance yin and yang within the body, which in turn guaranteed the flow of energy. As climate conditions influence the balance of yin and yang within the body, food advice during travel depended on the weather:

> During the course of your travel you will have to endure both hot and cold weather, but particular care should be taken in the summer for the heat weakens the internal organs and impedes digestion. And so it is best not to eat a lot of unfamiliar types of fish, fowl, or shellfish, as well as bamboo shoots, mushrooms, melons, pounded rice, and kowameshi. In summer, food poisoning, accompanied by vomiting

and diarrhea, may be a problem. Indeed, summer is the season in which you must take most care.[65]

A major concern was also the consumption of water. Generally, it was assumed that drinking water during travel, which was different from the water one was accustomed to, could lead to problems with digestion, increased blood flow to the head, and even result in pox.[66] Furthermore, food advice for travelers followed instructions of health manuals concerning the amount of food to be consumed and especially paid attention to the consumption of alcohol, for example, warned the traveler not to drink alcohol before taking a hot bath. In general, alcohol was considered to have medical properties and mild consumption of *shōchū* (strong spirit) or *awamori* (strong liquor from Okinawa) is recommended in the *Yōjōkun* and also in the *Ryokō Yōshinshū* to "relieve heart stroke".[67] The main worry concerning the consumption of alcohol in travel writings was the negative effect on human behavior and social harmony, and Ekiken[68] as well as travel guides warned of the inherent danger due to quarrels in inns as a result of a loss of control after the consumption of alcohol.[69]

The experienced traveler would therefore be prepared for different weather conditions and equipped with basic medicine that could, for example, be carried in a little box (*inrō*) attached to the waist belt. Different herbs, pills, and potions could come in quite handy as Hishiya Heishichi records in his *Tsukushi kiko* (*Account of a Journey to Tsukushi*) in 1802:

> Today I met two sixteen and seventeen-year-old girls in the mountains half a ri from Katsurano. They were on a *nukemairi* pilgrimage to Ise. One of them had sunstroke and lost her voice; she was pale and seemed in pain. I took pity on her and gave her some medicine before I continued on my way.[70]

Also, Bashō felt the power of the sun during his travels through the North:

Akaaka to	Red red
Hi wa tsurenaku mo	The relentless sun
Aki no kaze	Despite the autumn wind[71]

Herbs and medicine could be bought in *kampō* pharmacies along the roads, but charlatans also tried to sell their potions against all kinds of ailments on the roads, and the *Ryokō Yōshinshū* therefore warns to not trust fellow travelers selling panacea, but to buy medicine in the local pharmacies.[72] Herbs and medical potions mentioned to be useful in a medicine chest for travelers included ginger, pepper, crow dipper (*pinellia ternata*), angelica, and blue morning glory (*ipomoea indica*).[73] Especially pepper was considered to be helpful against different disorders and maladies that could affect the travelers including poisonous insects, different toxins, blisters, or dampness in the mountains.

Travel guidebooks as well as household books reflect the worries of travelers concerning their health and their need for keeping their bodies fit for traveling. Accordingly, all eventualities of bodily inconveniences on the road are covered: advice on how to not become seasick or sick in a palanquin, protection against wild animals and spirits, and how to relief pain and fatigue, to name but a few. The *Shūi chie no umi* (*Addendum to the Sea of Wisdom*) from 1788 mentions a potion made from dried orange skin, sweet vetch, kuth root, and warm salty water, which was supposed to help in cases of side ache and difficulties to get up.[74] Special attention is given to footwear, feet, and legs. While sandals made from whale fin are recommended as they last for a hundred miles, warrior sandals, made from leaves and ginger are considered to last only for 20 miles.[75] But even good sandals could result in blisters and the *Ryokō Yōshinshū* accordingly suggests:

> People unaccustomed to traveling tire or get blisters on their feet because they do not know how to wear their straw sandals properly. Get a good pair of sandals and pound them to soften the material. Don't be in a hurry when you put them on, otherwise the material may stretch or bunch up. Hot and dry feet will hurt and form blisters, so be sure to take proper rest breaks from time to time, removing your sandals and cooling your feet.[76]

Yet illness could not always be cured and pain not always be relieved during travel. Interrupting a journey meant to surrender to one's body and separation from fellow travelers. This also happened to Bashō during his journey to the north, when his companion and disciple Kawai Sora was afflicted by stomach troubles and decided to part:

Yuki yukite	I wander and wander
Taore fusu tomo	And even if I lay down
Hagi no hana	Under the bush clover.[77]

Notes

1 Shokyū-ni, quoted in Keiko Shiba, *Literary Creations on the Road. Women's Travel Diaries in Early Modern Japan* (Lanham: University Press of America, 2012), p. 79.

2 The *Oku no hosomichi* will be quoted according to Shōichirō Sugiura, Saburō Miyamoto, and Kyōshi Ogino, eds., *Bashō bunshū. Nihon koten bungaku taikei* 46 (Tokyo Iwanami shoten 1989), p. 85. For a philological and cultural interpretation and analysis of the *Oku no hosomichi* see esp. Geza Dombrady,

Bashō. Auf schmalen Pfaden durchs Hinterland (Mainz: Dietrich'sche Verlagsbuchhandlung, 1985).

3 Bashō, quoted in Sugiura, Miyamoto, and Ogino, *Bashō*, p. 87; see also Dombrady, *Bashō*, p. 171.

4 The phrase *bari suru* (to pee) also includes a clever reference to the place name of the barrier, which is Shitomae ("Before the urine"). The name refers to a legend connected to the famous warrior Minamoto Yoshitsune. When he and his wife passed this area when fleeing from his brother (Yoritomo), their newborn son peed for the first time (Dombrady, *Bashō*, pp. 170–71).

5 Philipp Franz von Siebold remarks that in no other Asian country is traveling as daily an occurrence as it is in Japan (Herbert Scurla, *Reisen in Nippon. Berichte deutscher Forscher des 17. und 19. Jahrhunderts aus Japan* (Berlin: Verlag der Nation 1990), p. 379). For different genres in travel writing see, for example, Herbert Plutschow, "What Pre-Modern Japanese Travel Writing Tell Us", *Review of Japanese Culture and Society*, vol. 19, *Aspects of Classical Japanese Travel Writing* (2007): 132–48; Jilly Traganou, *The Tōkaidō Road. Traveling and Representation in Edo and Meiji Japan* (London: Routledge Curzon, 2004), pp. 92–144; and Robert F. Wittkamp, *Japans frühmoderne Reiseliteratur. Leben und Werk von Sugae Masumi (1754–1829)* (Hamburg: Gesellschaft für Natur- und Völkerkunde Ostasiens e.V., 2001), pp. 13–62.

6 Concerning the complex relationship between public and private, fiction and non-fiction in Japanese travel literature see esp. Plutschow, "Pre-Modern Japanese Travel Writing", p. 132. For a general introduction to questions of how the frames of travel will shape not only the language but also the structure of narratives, see Alasdair Pettinger and Tim Youngs, eds., *The Routledge Research Companion to Travel Writing* (New York: Routledge 2019), as well as Ansgar Nünning, "Zur mehrfachen Präfiguration/Prämediation der Wirklichkeitsdarstellung im Reisebericht. Grundzüge einer narratologischen Theorie, Typologie und Poetik der Reiseliteratur", in Marion Gymnich et al., eds., *Point of Arrival: Travels in Time, Space, and Self. Zielpunkte: Unterwegs in Zeit, Raum und Selbst* (Tübingen: Francke, 2008), pp. 11–32; Uta Schaffers, *Konstruktionen der Fremde: Erfahren, verschriftlicht und erlesen am Beispiel Japan* (Berlin, New York: De Gruyter, 2006); and Uta Schaffers, Stefan Neuhaus, and Hajo Diekmannshenke, eds., *(Off) the Beaten Track? Normierungen und Kanonisierungen des Reisens* (Würzburg: Königshausen & Neumann, 2018).

7 Plutschow, "Pre-Modern Japanese Travel Writing", p. 136.

8 Plutschow, "Pre-Modern Japanese Travel Writing", p. 136.

9 Koshōken, quoted in Herbert Plutschow, *A Reader in Edo Period Travel* (Kent: Global Oriental, 2006), p. 94.

10 Timon Screech, *Sex and the Floating World. Erotic Images in Japan 1700–1820* (Honolulu: University Press of Hawai'i, 1999), p. 216; see also Timon Screech, *The Western Scientific Gaze and Popular Imagery in Later Edo-period Japan: The Lens within the Heart* (London: Cambridge University Press, 1996).

11 Manfred Pfister, "Intertextuelles Reisen oder: Der Reisebericht als Intertext", in Herbert Foltinek, Wolfgang Riehle, and Waldemar Zacharasiewicz, eds., *'Tales and Their Telling Difference': Zur Theorie und Geschichte der Narrativik. Festschrift zum Geburtstag von Franz K. Stanzel* (Heidelberg: Universitätsverlag Winter GmbH 1993), pp. 109–32 (at p. 119).

12 In the introduction to the *Oku no hosomichi*, Bashō also includes the topos of "suffering" in a classical sense: "Many poets that lived before us died on their peregrination" (Bashō, quoted in Sugiura, Miyamoto, and Ogino, *Bashō*, p. 70).

13 Pettinger and Youngs, *Companion*, p. 3.
14 See Wittkamp, *Reiseliteratur*, p. 44.
15 Shogo Tanaka, "The Notion of Embodied Knowledge", in Paul Stenner, eds., *Theoretical Psychology: Global Transformations and Challenges* (Vaughan: Captus Press 2011), pp. 149–57 (at p. 149).
16 See in this context esp. Nünning, *Präfiguration*, p. 14; Jonathan Crary, *The Techniques of the Observer: On Vision and Modernity in the Nineteenth Century* (Cambridge, MA: MIT Press, 1990), p. 6; and Traganou, *Tōkaidō*, p. 2.
17 Constantine Nomikos Vaporis, *Tour of Duty. Samurai, Military Service in Edo, and the Culture of Early Modern Japan* (Honolulu: University of Hawai'i Press, 2008); Constantine Nomikos Vaporis, *Travel and the State in Early Modern Japan. Breaking Barriers* (Cambridge, MA: Harvard University Press, 1994); Traganou, *Tōkaidō*; Plutschow, *Reader*; Laura Nenzi, *Excursions in Identity. Travel and the Intersection of Place, Gender, and Status in Edo Japan* (Honolulu: University of Hawai'i Press, 2008); and Wittkamp, *Reiseliteratur*.
18 For *sankin kōtai* and the meaning of official traveling for the spread of central culture to the periphery (and vice versa) see Vaporis, *Tour*, pp. 205–36.
19 At the beginning of the nineteenth century, a total of 179 inns for *daimyō* (feudal lords) were in business. These inns were able to lodge up to 60 people, which meant that lower ranking members of a *daimyō's* travel group had to stay in less comfortable circumstances in common inns, temples, or even private houses. The entire economy of a village along the Tōkaidō could actually depend on *daimyō* travels (as the villages and nearby farmers usually also provided the food).
20 The permits and passports were issued by different authorities and even by shrines and temples. The two main characters Yaji and Kita in the fictious travel novel *Tōkaidōchū Hizakurige* (*A Shank's Mare*), written by Jippensha Ikku (1765–1831) and published between 1802 and 1822, for example, receive their passport from the family temple in exchange for paying "a hundred coppers" (*Hizakurige or Shank's Mare. Japan's Great Comic Novel of Travel and Ribaldry by Ikku Jippensha*, trans. Thomas Satchell (Tokyo, Rutland: Tuttle, 1960), p. 23). But there were means to avoid controls by traveling on small paths in the mountains instead of the main roads, paying a guide fee to avoid barriers; another often mention method was to change one's identity and to pretend to be a pilgrim or priest (Nenzi, *Excursions*, pp. 75–76; pp. 90–91). Writer and ethnologist Sugae Masumi (1754–1829) – while complaining about travel restrictions – travels as an impostor and succeeds in passing controls more easily by pretending to be a *yamabushi* pilgrim (Plutschow, *Reader*, p. 148).
21 See Shiba, *Creations*, pp. 1–69.
22 Franziska Ehmcke, "Reisefieber in der Edo-Zeit", in Franziska Ehmcke and Masako Shōno-Sladek, eds., *Facetten der städtischen Bürgerkultur Japans vom 17. – 19. Jahrhundert* (München: Iudicium, 1994), pp. 55–70 (at pp. 62–63).
23 James H. Foard, "Buddhism and Tradition in Japanese Pilgrimage", *Journal of Asian Studies*, 41, no. 2 (1982): 231–51 (at p. 239).
24 Nenzi, *Excursions*, p. 46.
25 Inoue Tsū-jo, quoted in Shiba, *Creations*, p. 73.
26 Vaporis, *Tour*, pp. 12–13, 155–74. See also Nenzi, *Excursions*, p. 70–75. Similarly, Furukawa Koshōken commented on that topic in his diary while traveling to the Northeast and Ezo on a shogunal inspection tour (quoted in Plutschow, *Reader*, p. 103).
27 Shiba, *Creations*, p. 71.

28 For hair symbolism in (early modern) Japan see esp. Andreas Niehaus, "'Bird Never Nest in Bare Tree.' – Überlegungen zu einer praxisorientierten Kulturwissenschaft am Beispiel des Haare Machens in der Japanischen Kultur", in Andreas Niehaus and Chantal Weber, eds., *Reisen, Dialoge, Begegnungen. Festschrift für Franziska Ehmcke* (Münster: LIT Verlag, 2012), pp. 107–19.

29 Tsuchiya, quoted in Shiba, *Creations*, p. 78.

30 Tsuchiya, quoted in Shiba, *Creations*, p. 78.

31 Yasumi, quoted in Hartmut O. Rotermund, *Säcke der Weisheit und Meere des Wissens. Alte japanische Hausbücher – Ein kulturgeschichtliches Lesebuch* (München: Iudicium 2010), p. 317. Further references to this edition are included parenthetically in the text.

32 Asai, quoted in Ekkehard May, *Das Tōkaidō meishoki von Asai Ryōi. Ein Beitrag zu einem neuen Literaturgenre der frühen Edo-Zeit* (Wiesbaden: Harrassowitz, 1973), p. 139.

33 Norinaga, quoted in Plutschow, *Reader*, p. 58. These comments about the hardships and pain during travels are made by Motoori Norinaga (1730–1801), the famous scholar of the national studies (*kokugaku*) movement, in his *Sugagasa no nikki* (*Sugagasa Diary*), an account of a journey undertaken in 1722 from Matsuzaka to Yoshino.

34 Inoue Tsū-jo, quoted in Shiba, *Creations*, p. 73.

35 The technique of riding on a rented horse for commoners also differed from the one of the warrior class that used saddles. Commoners generally sat with crossed legs on a sort of platform or even in baskets on both sides of the horse's flanks.

36 Siebold, quoted in Scurla, *Nippon*, p. 381.

37 Ikku, *Tōkaidō*, quoted in Satchell, *Hizakurige*, p. 101.

38 Dombrady, *Bashō*, pp. 31–36.

39 On human excrements as fertilizers in the Edo-period see Anne Walthall, "Village Networks. Sōdai and the Sale of Edo Nightsoil", *Monumenta Nipponica*, 43, no. 3 (1988): 279–303. On smell, smellscapes, and excrements see also Clare Brant, "An Introduction to Smell Studies", in Pettinger and Youngs, eds., *Companion to Travel Writing*, pp. 249–61.

40 Asai, quoted in Plutschow, *Reader*, p. 7.

41 Plutschow, *Reader*, p. 7. See also May, *Tōkaidō meishoki*, p. 139 and p. 230.

42 Asai, quoted in May, *Tōkaidō meishoki*, p. 139, transl. by Andreas Niehaus. Asai is describing the Hakone climb on the Tōkaidō road, a road that was well maintained and for which Kaempfer noted that it is "in some seasons crowded as the alleys of European populous cities" (Kaempfer, quoted in Scurla, *Nippon*, p. 101). However, less well traveled or maintained roads would mean that travelers could fall and sometimes had to climb up and down on hands and feet (see, for example, Sugae Masumi, quoted in Wittkamp, *Reiseliteratur*, p. 245; and Bashō, quoted in Dombrady, *Bashō*, p. 175–77).

43 See Sugae, quoted in Plutschow, *Reader*, p. 137; and Wittkamp, *Reiseliteratur*, pp. 206–207.

44 Indeed, traveling was not as safe as it is today. Robbers, illness, and the weather were real threats to a traveler's life and therefore became topics in the diaries. The danger of being robbed significantly increased in periods of famine, and the physician Tachibana Nankei (1752–1805) comments in his *Saiyūki* (*Journey to the West*) on the situation in Kyushu during the winter of 1782: "Usually, Kyushu has a rich rice harvest and rice is much cheaper here than in Kyoto, but the situation then was unspeakably bad. Towards the end of winter, gangs of robbers rampaged through the provinces of Kyushu, and it was very dangerous to travel.

It happened that someone shot a traveler to death from behind a tree and stole his clothes and money" (Nankei, quoted in Plutschow, *Reader*, p. 80). See also Harold Bolitho, "Traveler's Tales. Three Eighteenth-Century Travel Journals", *Harvard Journal of Asiatic Studies*, 50, no. 2 (1990): 485–504 (at p. 494).

45 See, for example Rotermund, *Weisheit*, pp. 325–26; and Constantine Nomikos Vaporis, "Caveat Viator. Advice to Travelers in the Edo Period", *Monumenta Nipponica*, 44, no. 4 (1989): 471–83.

46 See Vaporis, "Caveat Viator", pp. 478–79.

47 In times of famine, travelers could also not anymore rely on the hospitality of locals, and Tachibana Nankei, for example, is at a certain moment only invited to stay in a private home after he assured to have food with him. This is different for travelers on official state business or lords on their tour of duty.

48 Already contemporary vernacular health manuals consider sleep an important factor in health care and give instructions concerning where, when, and how one should lie down (head to the north, sleep on one side, mouth closed. See Kaibara Ekiken, *Yōjōkun. Wazoku dôjiku,* Ken Ishikawa, ed. (Tokyo: Iwanami 2001), p. 38, pp. 44–46, pp. 101–103; also in Andreas Niehaus and Julian Braun, *Kaibara Ekiken (1630–1714). Regeln zur Lebenspflege (Yōjōkun)* (München: Iudicium 2010), pp. 42–43, pp. 49–51, pp. 119–121.

49 Falling asleep, or better not falling asleep, in inns or houses of hosts is, for example, commented on by Ueda Akinari (1734–1809) in his *Akiyama no ki* (*Record of a Journey to Akiyama*) where he complains that he and his wife could not sleep due to a party next door (Plutschow, *Reader*, p. 71). In Sugae Masumi's *Ezo no teburi* diary, which records a journey in 1791, the narrator cannot fall asleep because of conversations between people staying in different rooms and then, after the lights went out, due to the sounds made by someone taking a night snack near the hearth.

50 Bedding was generally considered one of the sources for infections and illness.

51 Plutschow, *Reader*, p. 133.

52 Asai, quoted in May, *Tōkaidō meishoki*, p. 82, transl. by Andreas Niehaus.

53 Shokyū-ni, quoted in Shiba, *Creations*, p. 79.

54 Rotermund, *Weisheit*, pp. 309–44; and Vaporis, "Caveat Viator".

55 Andreas Niehaus, "'They Should Be Called Gluttons and Be Despised': Food, Body and Ideology in Kaibara Ekiken's (Yōjōkun)", in Andreas Niehaus and Tine Walravens, eds., *Feeding Japan. The Cultural and Political Issues of Dependency and Risk* (Cham: Palgrave Macmillan, 2017), pp. 19–51 (at pp. 36–38).

56 Niehaus, "Gluttons", pp. 36–46.

57 On health and travel during Edo-period and esp. on walking and health discourse (*yōjō shisō*) see Hironori Tanigama, "Edo jidai no tabi to kenkō", *Supōtsu kenkō kagaku kiyō*, 44 (2018): 27–37 (at pp. 28–29).

58 Concerning the *Yōjōkun* as a guide to living a healthy life based on Confucian and Taoist thought as well as *kampō* (Chinese) medicine see the introduction to the German translation of the *Yōjōkun* by Niehaus and Braun, *Yōjōkun*, pp. 7–23.

59 Kaibara, *Yōjōkun*, p. 29, pp. 118–19; and Niehaus and Braun, *Yōjōkun*, pp. 32–33, pp. 139–40.

60 Quoted in Scurla, *Nippon*, p. 385.

61 Kaibara *Yōjōkun*, p. 118; see also Niehaus and Braun, *Yōjōkun*, p. 139.

62 Quoted in Rotermund, *Weisheit*, p. 316.

63 *Banpō chie-bukuro*, quoted in Rotermund, *Weisheit*, p. 320.

64 *Minka nichiyō denkahō*, quoted in Rotermund, *Weisheit*, p. 330.

65 Yasumi, quoted in Vaporis "Caveat Viator", p. 472.

66 Rotermund, *Weisheit*, pp. 330–31.

67 Yasumi, quoted in Vaporis, "Caveat Viator", p. 472.

68 Kaibara, *Yōjōkun*, pp. 91–94; and Niehaus and Braun, *Yōjōkun*, pp. 106–10.

69 See Rotermund, *Weisheit*, pp. 325–26; and Vapori, "Caveat Viator", p. 472. Alcohol consumption as well as prostitution was also identified as a major concern in the behavior of warriors that accompany their lord on their tour of duty (Vaporis, *Tour*, p. 22).

70 Hishiya quoted in Plutschow, *Reader*, p. 257. On pilgrimage see also Susanne Formanek, "Pilgrimage in the Edo-Period: Forerunner of Modern Domestic Tourism? The Example of the Pilgrimage to Mount Tateyama", in Sepp Linhart and Sabine Frühstück, eds., *The Culture of Japan as Seen through Its Leisure* (New York: State University of New York Press, 1998), pp. 165–93.

71 Bashō quoted in Sugiura, *Bashō*, p. 93. See also Dombrady, *Bashō*, p. 233, and pp. 313–14.

72 Rotermund, *Weisheit*, p. 325.

73 Rotermund, *Weisheit*, pp. 317–23; and Vaporis, "Caveat Viator", p. 473.

74 Rotermund, *Weisheit*, p. 318.

75 Rotermund, *Weisheit*, p. 320.

76 Yasumi, quoted in Vaporis, "Caveat Viator", p. 483.

77 Sora, quoted in Sugiura, *Bashō*, p. 95.

9 "The 'Food Question' Is Said to Be the Most Important One for All Travelers"

Eating in Travel Writing

Uta Schaffers

9.1 Introduction

Among all dissuasions and good advice that Isabella Bird (1831–1904) receives from fellow adventurers, experts, friends, and acquaintances regarding her tour along the *Unbeaten Tracks of Japan* in 1878, she states in one of her letters that:

> The 'Food Question' is said to be the most important one for all travelers and it is discussed continually with startling earnestness, not alone as regards my tour. However apathetic people are on other subjects, the mere mention of this one rouses them into interest. All have suffered or may suffer, and every one wishes to impart his own experience or to learn from that of others. Foreign ministers, professors, missionaries, merchants – all discuss it with becoming gravity as a question of life and death, which by many it is supposed to be.[1]

Considering that in the late nineteenth century it was still very unlikely for a woman to travel long distances without any other purpose than to travel and to write about it, it seems odd that of all things, the "Food Question" is ascribed such "gravity". One would expect her gender to be more of an issue; after all, she was aiming for the 'unbeaten tracks' in the north of Japan (Hokkaido) by foot and on horseback without a suitable male chaperone. Although Bird showed an "insistence of outward propriety"[2] – not the least in her travel outfits – and her demeanor could be called feminine by the norms of her time, through her travels and by undergoing the physical act of traveling, she claimed a subject position that had traditionally been embodied by men. Therefore, like many other traveling women of that period, she developed at least some discursive and practical strategies of justifying[3] as well as more or less playful forms of masquerade, camouflage, and gender performances.[4] Still, as "the first European lady who had been seen in several districts" (Bird xix), Bird became a matter of interest and irritation among the Japanese:

DOI: 10.4324/9781003331803-12

> I took my lunch – a wretched meal of a tasteless white curd made from beans, with some condensed milk added to it – in a yard, and the people crowded in hundreds to the gate, and those behind, being unable to see me, got ladders and climbed on the adjacent roofs, where they remained till one of the roofs gave way with a loud crash, and precipitated about fifty men, women, and children into the room below which fortunately was vacant. [...]. The Transport Agent begged them to go away, but they said they might never see such a sight again! One old peasant said he would go away if he were told whether 'the sight' were a man or a woman, and, on the agent asking if that were any business of his, he said he should like to tell at home what he had seen [...].

> (Bird 146)

As much as Japan and the Japanese were a "sight" – an *Eräugnis* – for the Western travelers in the years following the forced 'opening' of Japan in 1853–54,[5] the travelers themselves were a sight too, and as such they became narrative material: "he should like to tell at home what he had seen". There is consensus that visual perception ("those behind, being unable to see me"; "might never see such a sight again"; "what he had seen") is the most important – and a highly controversial[6] – sense when it comes to travel and travel writing.[7] To have seen something with one's own eyes or to read that someone has seen something is the 'founding myth' of travel writing; these are the magical words that open the (perceptual and perceived) field or space, and the world seems to be given to the eyes of travelers and readers.[8] But travelers not only see: they also hear and touch, they smell and taste, they are in a given environment with all their senses, as an organism, and they write about it.[9] The Japanese observed Bird while she took her lunch: "– a wretched meal of a tasteless white curd made from beans, with some condensed milk added to it –". The sensual impression (tastelessness), the evaluation ("wretched"), the ingredients (curd made from beans, condensed milk), and also, embedded in this description, the act of eating, the texture of the curd felt in the mouth, the chewing and swallowing (with little consent or pleasure), the whole process of incorporation linger between two dashes, even though all of this has no significant relevance for the actual event. But still it is there, and one could argue that it is even underlined by the dashes.

Exploring the 'Food Question' in travel texts reveals that it is indeed a "most important one to all travelers" (Bird 19). Based on the reflection of the ineluctable necessity and corporeality of eating, this article explores some of the intricate connections between traveling (bodies), food, eating, and the perception of the Other by focusing mainly on texts of Westerners traveling in Japan in the late nineteenth and early twentieth centuries.[10]

Traveling, food, and eating is a nexus that is inextricably bound up with telling and writing about it, regardless of whether the 'Other('s) Food'[11] is experienced as being exotic, palatable, or disgusting.[12] Sometimes, travelers even run out of food: "It is a long road and those who follow it must meet certain risks [...] let it come late October, or November, and the snow-storms block the heights, when wagons are light of provisions and the oxen lean, then will come a story."[13] Knowing that it is impossible to tell the whole story, I would like to shed light on two exemplary aspects that might reveal the relevance of food and eating for traveling bodies: national politics and disgust.

9.2 National Politics

Travelers are bodily experiencing organisms in interaction with an unconversant environment that could cause insecurities (see the introduction to this volume), and certainly food, this "liminal substance",[14] can become a source of (self-)observation and irritation. Deborah Lupton states that "Food and eating are central to [...] our experience of embodiment, or the ways that we live in and through our bodies" and not "simply matters of 'fuelling' ourselves [...;] bodily experiences and physical feelings are constructed or mediated by society and culture".[15] It is common knowledge and probably universal experience that food can occasionally provoke strong bodily reactions, varying from most pleasant to unpleasant (or totally unpleasant, as, for example, described by Fritz Blank in his article "Travelers' Diarrhea: The Science of 'Montezuma's Revenge'"[16]). Sometimes even the pure sight of foodstuff can induce repulsion, but also joyful anticipation, which are both part of the embodied dynamics of distance and proximity toward food. Such 'responses' of the body[17] might be perceived as being natural, but they are deeply interwoven with, for instance, ethical, religious, medical, and economic discourses, rules, and norms, as well as aesthetic considerations.[18] In that respect, discourses on the Other('s) food[19] and eating in travel texts are not merely articulations of a more or less 'authentic' pleasant or unpleasant bodily experience. They often exhibit a significant function in order to defame and denigrate a culture or nation while at the same time underlining one's 'own' culture or nation as superior, or to exoticize, even fetishize it. One might even argue that incorporating the Other('s) food is metaphorically and metonymically incorporating 'the Other'. Therefore, food and eating on the one hand can establish social relations between bodies,[20] cultures, and nations and on the other hand also constitute difference and reinforce discourses of the "*imaginary divides* ('us' and 'them')".[21] Keeping that in mind, one has to take into consideration that stories about food and eating in travel texts, be they on the more factual or fictional side of the continuum, are almost always political, and it can be observed that – in an act of discursive and/or performative 'blending' – food mobilizes strong emotions toward other cultures and/or nations.

Traveling bodies are eating bodies: we need to eat in order to travel (in order to live), and sometimes we also travel to eat.[22] Unlike today's dimension of globalization and mobility – including the mobility of food in the sense of creating global and transnational "foodscapes",[23] independent from geography, cultures, and nations – in the nineteenth and early twentieth centuries a journey to Japan came with a departure from one's own food, eating heritage, and habits. Back then, Japan was certainly not considered to be a hot spot for culinary tourism. Disregarding all other obstacles, one could face on such a tour (mosquitoes, fleas, and bedbugs seemed to have been a question of some gravity too), Western travelers had to learn "not to be particular!" (Bird 105).[24] They had to prepare themselves to encounter a diet that "fail[s] to satisfy European cravings", as Basil Hall Chamberlain, one of the leading British Japanologists of his time, puts it in *Things Japanese* (1890), an encyclopedia *For the Use of Travellers and Others*:

> Imagine a diet without meat, without milk, without bread, without butter, without jam, without coffee, without salad or any sufficient quantity of nicely cooked vegetables, without pudding of any sort, without stewed fruit and with comparatively little fresh fruit, – the European vegetarian will find almost as much difficulty in making anything out of it as the ordinary meat-eater. [...] The food is clean, admirably free from grease, often pretty to look at. But try to live on it – no! [...] A foreigner forced by circumstances to rely on a Japanese diet should, say the doctors, devote his attention to beans.[25]

Nowadays 'Japanese cuisine' is often considered to be remarkably healthy and pure (enthusiasts of Japanese food would still agree that it is "clean, admirably free from grease, often pretty to look at"), and it has become a distinctive marker of good 'taste'.[26] Reading the chapter "Food" in Chamberlain's book, however, gives the impression of a diet that lacks almost everything that was at that time considered to be nutritious, healthy, tasty, and even essential. Japanese food is presented as a cuisine of deficiencies and shortcomings for every non-Japanese (and for the Japanese themselves, one could imagine). This perception and evaluation has various reasons: for example, the traveler's eating habits and gustation had not yet been 'globalized', maybe they simply had bad luck, or they misunderstood or underestimated aspects such as the regional, social, and economic status of their hosts, let alone religious or dietary rules based on different medical and ideological traditions (like for example *kampō* medicine or Confucianism). Apart from that, one should not forget that Japanese cuisine, like any other, has undergone some transformations since premodern times.[27]

But there is more to that. Because "[f]ood is a handy, readily available place to look for evidence of Otherness",[28] and because it is so substantial, it provides a perfect field for discourses of power and establishing hierarchies. For the time period focused on in this article, the power relations between Japan (East Asia) and the West – as well as *among* the East Asian and Western powers – were forcefully 'negotiated' in fields like politics, economics, science, education, and other institutions,[29] including military conflicts, such as the Russo-Japanese War of 1904–05. In this historical context, for Western travelers in Japan, it was self-evident to face Japan (and Japanese food) with an imperialistic and colonializing attitude[30] – although Japan was never colonialized by the West – as well as an almost imperturbable belief in the superiority of their own culture and/or nation.

The concept of 'national cuisines' is often part of the joy of traveling, but it comes with some significant issues. It relies on concepts of tradition, purity, and authenticity and conveys the idea of a meal in a snowball: for generations, the same foodstuff, homegrown vegetables, crops, and animals, are harvested and slaughtered in one's own country, prepared in the same old ways (by the same mothers and housewives), eaten with the same manners and practices – under glass and protected from any 'contagion' from the outside world.[31] In fact, "[n]ational cuisines are in a process of constant reinvention",[32] influenced by such things as intercultural contacts, weather conditions, individual taste, medicine, fashion, economics, political discourses, and not least by matters of social and economic class. But like many (invented) traditions,[33] food as well as the practices and rituals evolving around the consumption of food strongly participate in the nation-building process and in (national) identity politics,[34] which can perfectly be observed in times of making nations great again.[35] David Bell and Gill Valentine state: "If, as Benedict Anderson (1983) has famously proclaimed, the nation is an 'imagined community', then the nation's diet is a feast of imagined commensality."[36] The longing to become part of this imagined community and commensality (with a return ticket) is expressed in the somewhat nostalgic search of tourists to find an 'authentic' cuisine at their destination, a practice often declared as 'going native'. As illustrated, the discourses around Japanese cuisine had been different about 120 years earlier and did not encourage travelers to 'go native' in that respect (and one could certainly observe the same about the reputation of Western food in Japan).

Travel literature has different functions, such as potentially serving as preparatory reading for one's own journeys. The specific explicit and implicit intertextuality of travel books on Japan[37] reveals that these books were received in a transnational context. Arthur Neustadt, a German travel author who published a book called *Japanische Reisebriefe* (1913),[38] refers to writers such as Bird and Chamberlain. In 1911, when Neustadt traveled in Japan, hotels had started to provide special food for foreigners that would fit their wishes and needs, even in the form of picnic baskets. In one of his letters, Neustadt writes about an Englishman who "strangely enough" did not prepare himself for the "primitive guesthouses" of the "interior" Japan with a reading of the above-mentioned famous English travelers and travel writers and failed to bring any suitable food – a clear sign of an unfortunate lack of what was considered vital information for Western travelers in Japan. When Neustadt entered the inn, he observed that this Englishman, as a result of his neglect, "struggled to choke Japanese food" with a "desperate effort" while using the "little chopsticks".[39]

Eating while traveling has always been an opportunity for encounters on various levels, and on some occasions, they were (and are) explicitly and officially political.[40] As such, they harbor some pitfalls, and for the sake of the nation and international politics, it was considered necessary to overcome the routine and the immediateness of the body and bodily reactions. In her chapter "Eating for 'Civilization and Enlightenment'", Katarzyna Cwiertka describes the effort of the Japanese court in 1871 to "become acquainted with Western ways, including Western food and Western dining etiquette".[41] Etiquette refers to a bodily performance, to eating habits, norms, and rules as well as to body techniques (Mauss[42]) and eating gestures (Simmel[43]) that arise from such things as the use of specific artifacts. But it also includes bodily reactions that could cause diplomatic irritations: the sensory system and the whole organism have to get used to different views, tastes, textures, and internal effects of the Other('s) food. Bird wrote quite frankly: "The fact is that, [...] food must be taken, as the fishy and vegetable abominations known as 'Japanese food' can only be swallowed and digested by a few, and that after long practice" (19).[44]

As mentioned before, it was a result of the growing tourism and the intercultural exchange between Japan and the West that more and more suitable food services for foreigners were established in Japan in the early twentieth century. But in the 1860s, when official Japanese delegations traveled to Europe and the United States, they had to bring their own food, their own cooking and eating utensils, and their own cooks. Regarding the long distances, the provisions did not last long, and as a consequence, they had to 'go native' or to live with the efforts of their hosts to provide appropriate meals. The letters and reports tell vivid stories of distress endured by the Japanese, such as trying to swallow rice cooked in milk or with sugar, as was offered during a stay in Philadelphia in 1860.[45] Sometimes it was "well beyond the power of my pen to describe what we, the Japanese, suffer on our journey to a foreign country".[46] The importance of food and eating for the well-being of a person in its entirety could conflict severely with the well-being of the nation and national politics, and the settlement of these conflicts did not always favor one's personal well-being: traveling

in Europe in 1862, the participants of a Japanese mission had serious problems getting used to the served meat, which they considered to be greasy.[47]
In France, they had access to slices of raw fish and tried to prepare it as
sashimi. But when they read in an English newspaper that, regarding this
food choice, they were compared with "the natives of South America" –
which basically meant being "recognized"[48] and marked as 'the uncivilized
Other', and not being recognized and accepted as a serious player within
the power relations of the 'civilized world'[49] – they abstained from the
consumption of raw fish "for the sake of national dignity"[50] and chose to
mistreat their bodies on this diplomatic journey.

The politics and discourses of food and eating are multifaceted, and I
would like to illustrate the 'intersection' of these by briefly focusing on the
'Meat Question'. Considering that "[t]he issue of meat-eating is one of the
most contentious aspects in Japanese food-history",[51] I want to point out
that, after a long period of restrictions on meat-eating (that had been followed more or less strictly), the consumption of meat became part of the
official Japanese policy in the period of modernization, but it took some
time to provide enough meat and to develop adequate ways and variations
of preparation. Western travelers were aware of 'a meat-situation' in Japan
during that period. Being prepared through travel texts and other sources,
they brought meat in various forms, like tinned meat, meat stock (*Liebigs
Fleischextrakt*), or even "meat lozenges", as did Bird, who opened her
"last resort, a box of Brand's meat lozenges, and found them a mass of
mouldiness" (196). Meat lozenges were perfect for travelers (maybe not so
perfect in terms of aesthetics, let alone taste) who had to eat on the move
in a country where a shortage of meat was expected.[52] The possibilities of
getting meat in restaurants in Japan were yet welcomed in a quite ambivalent manner, because there were some concerns about the appropriation
of Western food culture in Japan. In *Things Japanese*, Chamberlain judges
these restaurants quite harshly: "Unfortunately, third-rate Anglo-Saxon
influence has had the upper hand here, with the result that the central
idea of the Japano-European cuisine takes consistency in slabs of tough
beefsteak anointed with mustard and spurious Worcestershire sauce."[53]
Traces of discourses on class,[54] relating to his own social and national
background – third rate means on the one hand 'low of quality', but also
implies class, as lower-class restaurants with inferior quality of meat –
overlap with what might be doubts about the overall 'ability' of Japanese
to prepare foodstuff that was (and maybe still is) considered to be the most
important part of the national British cuisine: beef.

Eating 'proper' meat was (and is) an important marker for standards
of living, power, and strength, and in terms of politics some Westerners
ventured the idea that "meat-eaters dominated world politics".[55] It does
not come as a surprise that meat-eating is significantly gendered[56] and a

strong sign of masculinity. In travel texts on Japan, these implications of consuming meat were blended with power discourses on the East and the West: the German author Max Dauthendey (1867–1918) traveled around the world with Thomas Cook, and Japan was one of his destinations (from 23 April until 24 May 1906). He processed his experiences in various genres,[57] and in 1910 he published *Die geflügelte Erde*, a somewhat challenging versed travel book. In one of the chapters on Japan, the 'travel writer' ["Reiseschreiber"[58]] enters a hotel and gets a meal, served by one "tiny landlord" and ten "tiny handmaids"[59] (like Snow White's seven dwarfs beyond the seven mountains). In the following, the 'travel writer' refers to himself – figuring a 'Japanese' perspective – as a "clumsy European pig", who enters the room with boots and cuts huge amounts of meat to pieces with a knife and in a powerful manner; it does not seem to be entirely certain that he will not include the "tiny Japanese" in his horrific barbarian meal.[60] This awkward, staged change of perspective participates in fact in a very common imperialistic and racist discourse: based on various strategies, it was a tradition to turn Japan into some kind of Lilliput and Blefuscu from Jonathan Swift's *Gulliver's Travels* (1726), and to diminish and downsize the Japanese (bodies)[61] while the Westerners were turned into a huge Gulliver.[62] In a letter to his wife (29 April 1906), Dauthendey took this even a step further by characterizing Japanese sweets as "ethereous", maybe able to tempt the palate of an "embryo in the womb", but "for our tongues they are like filtered and distilled air".[63] One has to take into account that in 1906, when Dauthendey traveled in Japan, the Western powers were still struggling with the fact that Japan had defeated Russia in the Russo-Japanese War of 1904–05. Dauthendey himself witnessed in Tokyo the biggest and most spectacular exhibition and parade that displayed Japan's national military power, including the captured weapons and military equipment of the Russians.[64] The scene above with the meat-devouring, masculine Western giant, and the tiny Japanese landlord next to the tiny Japanese women (later referred to as "kittens") attains a special flavor against this background: one could read it as an attempt to process a serious mortification (not the least by claiming space) and to reestablish power hierarchies by means of food, eating, gendering, rhetoric, and literature.

9.3   Disgust

Another example of the lack of soothing comments on Japanese cuisine during the late nineteenth and early twentieth centuries can be found in the *Handbook for Travellers in Japan* (1891), in which Chamberlain and W.B. Mason state: "Many who view Japanese food hopefully from a distance, have found their spirits sink and their tempers embittered when brought face to face with its unsatisfying actuality."[65] In this quote, the overall visual perception is crucial anew: "view"; "brought face to face". One of the remarkable features of vision is the possible span between proximity and distance between the object and the body. Chamberlain and Mason vote for distance, which ensures that direct bodily contact with the Other('s) food must not be feared. But if foodstuff that comes into view is about to be eaten, this distance has to be given up and an extreme form of proximity has to be tolerated: food "is placed in the mouth, chewed, tasted, swallowed and digested. Its solidity is thus broken down and rendered into fragments that both pass through, and become, the eater's body."[66] It is possible that foodstuff that looks tempting from a distance turns into something repulsive once it crosses the liminal zone: "These cakes are about the length of a person's hand, trough-shaped, and baked until they turn attractive yellow and red hues. They are a horror to eat though."[67] In their "Einführung in die Mundhöhle" ["Introduction to the Oral Cavity"], Hartmut Böhme and Beate Slominski state that the "oral cavity forms the contact zone between the interior world of the body and the exterior world of objects". Both continue:

> Substances from the external world must enter the body through the mouth [...]. This marks the beginning of a process [...] of internalization, through which the foreign, as far as it is agreeable to the palate, is transformed into 'one's own being' [...]. With the physiological nutrition the politics of assimilation and dissimilation, of inclusion and exclusion begin.[68]

All senses – sight, hearing, taste, smell, and touch – are involved in eating and in the dynamics of proximity and distance related to this practice. To *see* food is possible from quite a distance; to *hear* the sound it makes when (others) slurp it or chew it or to *smell* food, one has to be close enough: "the strong-smelling radish (*daikon*) which is as great a terror to the noses of most foreigners as European cheese is to the noses of most Japanese."[69] To *feel* its texture with the hand or in the mouth, to *taste* it, asks for intimate bodily contact, an utter proximity.

In terms of the last topic of this article, disgust, all senses are involved and all of them can provide data for disgust. It is remarkable that disgust provokes certain bodily reactions that turn the ultimate proximity (incorporation) into a powerful repulsion against what is touching us or in us:[70] "The fundamental schema of disgust is the experience of a nearness that is not wanted. An intrusive presence, smell or taste, is spontaneously assessed as contamination and forcibly distanced."[71] Although neurosciences and evolutionary biology argue that the cultural approach to disgust is misleading because disgust is a behavioral adaptation to avoid the ingestion of pathogens, it seems to be more fruitful to overcome such dichotomies and binary thinking. To recognize that disgust can be a behavioral adaptation

in order to protect the body from ingestion of pathogens does not necessarily mean it has no cultural implications or even basis:

> I was so far from well that I was obliged to sleep at the wretched vil-
> lage of Abukawa, in a loft alive with fleas, where the rice was too dirty
> to be eaten, and where the house-master's wife, who sat for an hour
> on my floor, was sorely afflicted with skin disease.
>
> (Bird 173)[72]

This whole scene is a blending of a dirty environment, animals that indicate
a lack of hygiene, a sickness of a 'liminal' organ – a skin disease is a visible
indicator of a possible contagion – and dirty food. All things considered,
it is almost impossible to imagine to eat *there*, *this*, prepared by *her* – or
to identify the one crucial element that might trigger disgust. The danger
of possible contamination with dirt and disease through contact with the
environment, the fleas, the foodstuff, and not least the woman overlaps,
and the revulsion[73] at all of these, including the woman and maybe even
her culture and nation, is readable.[74]

Through the reflection on a quote from Charles Darwin, who "felt utter
disgust at my food being touched by a naked savage, though his hands did
not appear dirty",[75] Sara Ahmed analyzes "The Performativity of Disgust"
and shows how the Other is "already seen as dirt"[76] even if he/she is not
dirty. Referring to the quote by Darwin, Ahmed hints at the nexus of
Otherness, nakedness, and the fear of contamination. She points out that
the encounter reflected by Darwin could only take place "given a certain
history whereby the mobility of white European bodies involves the transformation of native bodies into knowledge, property and commodity."[77]
Similar dynamics can be observed regarding the encounters of Bird and the
Ainu. In *Unbeaten Tracks of Japan*, she encountered inhabitants of Japan
who were in Japan itself considered to be 'savages'. Following the overall
perception and discourse up to a certain point, she wrote, "Ainos [sic], are
complete savages in everything but their disposition, which is said to be so
gentle and harmless that I may go among them with perfect safety" (217),
and she referred to them as "these simple savages" who "are children, [and]
as children to be judged" (242). In a somewhat ethnological manner – and
with the general attitude of superiority of a European traveling in Japan in
these years – Bird describes life, character, customs, social structures, practices, and not the least the bodies of the Ainu. In addition, the notorious
'Food Question' comes into sight. She reports on their diet, "which usually
consists of a stew of 'abominable things'" (268), and on their eating habits
and rituals. The 'travel writer' characterizes the stew as a "carnival" in the
sense of a kind of wild mixture of elements, "consisting of fresh bear's flesh
and sake, seaweed, mushrooms, and anything they can get, in fact, which

is not poisonous, mixing everything up together" (268). The cuisine of the Ainu is definitely not considered to be a 'national cuisine' even by the standards that were applied by Westerners to Japan in those days. Apart from the status of the Ainu as (noble) 'savages' without a nation but instead a "lonely Aino land" (234), the diet of the Ainu is perceived and conveyed not even as a cuisine, but as something without a systematic approach to food, defined ingredients, or a thoughtful way of preparing – in short, the food and eating culture of the Ainu is presented as a practice to fill the body and not as a practice to 'eat' chosen foodstuff in a 'civilized' way.

Now, how close does the traveler come to this other food and eating culture?

> Benri's two wives spent the early morning in the laborious operation of grinding millet into coarse flour, and before I departed, as their custom is, they made a paste of it, rolled it with their unclean fingers into well-shaped cakes, boiled them in the unwashed pot in which they make their stew of 'abominable things,' and presented them to me on a lacquer tray. They were distressed that I did not eat their food.
>
> (285)

Even though Bird is attracted to a lot of elements of the Ainu culture (not least to their bodies), and even though traveling bodies need to be fed, this is one of the moments where – in terms of food and eating – the oscillation between proximity and distance to the other culture leans toward distance: the Ainu community is not considered to be a culture or nation and Bird is not on a diplomatic journey, and she therefore does not feel any need to overcome her disgust 'for the sake of the nation'; even the women preparing the meal are not seen as individuals that should not be offended by refusing the offered provisions. In that respect, the traveling guest is not just refusing a meal that, for various reasons, seems inedible to her – by rejecting the food, this 'liminal substance', she draws a strict line between 'us' and 'them' and refuses community and commensality.

Notes

1 Isabella Bird, *Unbeaten Tracks in Japan: An Account of Travels in the Interior Including Visits to the Aborigines of Yezo and the Shrine of Nikko* [1885] (New York, Tokyo, Osaka, and London: ICG Muse, Inc., 2000), p. 19. Further references to this edition are included parenthetically in the text.
2 Evelyn Bach, "A Traveller in Skirts: Quest and Conquest in the Travel Narratives of Isabella Bird", *Canadian Review of Comparative Literature*, 22, no. 3–4 (1995): 587–600 (at p. 590).
3 Monica Anderson, *Women and the Politics of Travel, 1870–1914* (Madison: Fairleigh Dickinson University Press, 2006), pp. 82–83.
4 See also Sofie Decock's contribution in this volume: *Going Undercover? Female Bodies and Clothes under Scrutiny in Travel Literature.*

5 On 8 July 1853, the US commodore Matthew Perry arrived at the coastline of Japan with four ships; on 15 February 1854 he came back, this time with a bigger squadron of seven ships, and after a few weeks of negotiations, he forced the Japanese to open their harbors after about 200 years (since 1639) of policy seclusion to Western powers and demanded treaty relations (William G. Beasley, "The Foreign Threat and the Opening of the Ports", in Marius B. Jansen, ed., *The 19th Century. The Cambridge History of Japan*, vol. 5 (Ann Arbor: Cambridge University Press, 2007), pp. 259–398).

6 Mary Louise Pratt, *Imperial Eyes: Travel Writing and Transculturation* (New York: Routledge, 2008).

7 Giorgia Alù and Sarah Patricia Hill, "The Travelling Eye: Reading the Visual in Travel Narratives", *Studies in Travel Writing*, 22, no. 1 (2018): 1–15.

8 This idea of visual perception during travels comprises instruments and technological devices like photo cameras; Donna Haraway argues that these devices prolong the view and the gaze of the observer: "I would like to insist on the embodied nature of all vision, and so reclaim the sensory system that has been used to signify a leap out of the marked body and into a conquering gaze from nowhere" ("The Persistence of Vision", in Nicholas Mirzoeff, ed., *The Visual Cultural Reader* (London: Routledge, 2002), pp. 677–85 (at p. 677)). Although visual instruments do prolong the gaze of the person behind the camera, 'to see' is not like taking a photo: "Visual experiences do not present the seen in the way that a photograph does. [...] You aren't given the visual world all at once. You are *in* the world, and [...] enact your perceptual content, through the activity of skillful looking." (Alva Noë, *Action in Perception* (Cambridge, MA: MIT Press, 2006), p. 73).

9 As Alasdair Pettinger and Tim Youngs put it, senses "perform a mediating role", they "indicate and even shape how one travels" ("Introduction", in Alasdair Pettinger and Tim Youngs, eds., *The Routledge Research Companion to Travel Writing* (New York: Routledge, 2020), pp. 1–14 (at p. 3)). In their collection, the editors "suggest how critical approaches to the literary treatment of [senses] may enhance our understanding of travel writing" (at p. 8).

10 See Uta Schaffers, *Konstruktionen der Fremde. Erfahren, verschriftlicht und erlesen am Beispiel Japan* (Berlin, New York: De Gruyter, 2006); and Andrew Elliott, "British Travel Writing and the Japanese Interior, 1854–99", in Martin Farr and Xavier Guégan, eds., *The British Abroad since the Eighteenth Century. Travellers and Tourists*, vol. 1 (London: Palgrave Macmillan, 2013), pp. 197–217. With 'Westerners' or 'western part of the world', I refer to Western Europe and the United States as geographical spaces. Since 'geographical space' is a relational concept, the geographical outlines are variable. Also, the denomination 'West' is a discursive construct that contains dynamic historical, political, ideological, and power-related implications. I will also take some glimpses at the comments and texts of Japanese travelers in the Western world of the same time period in order to provide a – rather small – cross-cultural perspective when it comes to the perception of the 'Other('s) food'. Since I am not capable of reading Japanese, I have to rely on translated sources when it comes to Japanese texts. I would like to thank Andreas Niehaus for his advice and all the information he provided.

11 When the notion *the 'Other('s) Food'* is used in this article, it combines two meanings which both are an expression of a 'relation': (a) 'other food' in terms of unconversant food or food one is not necessarily used to, and (b) 'food of the Others'. I capitalize 'Other' to stress the fact that in this context food and the discourse on food of the 'others' are part of the process of 'Othering'.

12 Heidi Oberholtzer Lee states that the "descriptions of food and eating, taste and orality, production and consumption [...] can be read for their discourses of embodiment, incorporation, appetite, and desire" ("Tasting", in Pettinger and Youngs, eds., *Companion to Travel Writing*, pp. 236–49 (at p. 236)).

13 George R. Steward, *Ordeal by Hunger: The Story of the Donner Party* [1936] (Boston: Houghton Mifflin Company, 1992), quoted in Bill Schutt, *Cannibalism: A Perfectly Natural History* (Chapel Hill: Algonquin Books, 2017), p. 133.

14 David Bell and Gill Valentine, *Consuming Geographies* (London, New York: Routledge, 1997), p. 44.

15 Deborah Lupton, *Food, the Body and the Self* (London, Thousand Oaks: Sage Publications, 1996), p. 1. The research on food and eating is almost unmanageable, and the different angles and (inter-)disciplinary perspectives show a broad variety. The following volume provides an overview: Carol Counihan and Penny Van Esterik, eds., *Food and Culture: A Reader* (New York: Routledge, 2008).

16 Fritz Blank, "Travelers' Diarrhea: The Science of 'Montezuma's Revenge'", in Harlan Walker, ed., *Food on the Move: Proceedings of the Oxford Symposium on Food and Cookery 1996* (Devon: Prospect Books, 1997), pp. 38–43.

17 I would like to echo Mark Johnson, who suggests "at least five interwoven dimensions of human embodiment" ("What Makes a Body?", *The Journal of Speculative Philosophy*, 22, no. 3 (2008): 159–69 (at p. 164)): the body as biological organism, the ecological, phenomenological, social, and the cultural body. Johnson stresses that "[t]he human body has all five of these dimensions, and it cannot be reduced to any one (or two or three) of them" (at p. 166).

18 Zbigniew Białas, *The Body Wall: Somatics of Travelling and Discursive Practices* (Frankfurt: Peter Lang, 2006), p. 131: "Through taste we encounter aspects of the physical world and at the same time the language of taste is *par excellence* the language of aesthetics." Białas references Dabney Townsend, who states in *An Introduction of Aesthetics* that the term 'taste' "applied to the sense arising from direct contact with the tongue and it also applied to the additional sense that responded to the aesthetic properties of objects" ((Oxford: Blackwell, 1997), p. 15).

19 See, for example, Pasi Falk, *The Consuming Body* (Thousand Oaks: Sage Publications, 1994), pp. 79–81.

20 Emma-Jane Abbots and Anna Lavis, "Introduction: Contours of Eating – Mapping the Terrain of Body/Food Encounters", in Emma-Jane Abbots and Anna Lavis, eds., *Why We Eat, How We Eat: Contemporary Encounters between Foods and Bodies* (Farnham: Ashgate, 2013), pp. 1–12 (at p. 1).

21 Sara Ahmed, *Strange Encounters: Embodied Others in Post-Coloniality* (Abingdon: Routledge, 2000), p. 116, emphasis in text.

22 Sarah Gibson, "Food Mobilities: Traveling, Dwelling, and Eating Cultures", *Space and Culture*, 10, no. 1 (2007): 4–21 (at p. 16). See also Walker, ed., *Food on the Move*; Michael C. Hall et al., eds., *Food Tourism around the World. Development, Management and Markets* (Oxford: Butterworth-Heinemann, 2003); and Lucy M. Long, ed., *Culinary Tourism: Exploring the Other through Food* (Lexington: University of Kentucky Press, 2004).

23 Sylvia Ferrero, "*Comida sin Par*: Consumption of Mexican Food in Los Angeles: 'Foodscapes' in a Transnational Consumer Society", in Warren Belasco and Philip Scranton, eds., *Food Nations: Selling Taste in Consumer Societies* (New York: Routledge, 2002), pp. 194–219 (at p. 197).

24 See also Richard Hoskin, "'The Fishy and Vegetable Abominations Known as Japanese Food'", in Walker, ed., *Food on the Move*, pp. 127–37.

25 Basil Hall Chamberlain, *Things Japanese: Being Notes on Various Subjects Connected with Japan for the Use of Travellers and Others* (London: K. Paul, Trench, Trübner & Co., Ltd., 1890), p. 180.

26 Pierre Bourdieu, *Distinction: A Social Critique of the Judgement of Taste* (London: Routledge, 1984). See also Falk's chapter "Towards an Historical Anthropology of Taste" (*The Consuming Body*, pp. 68–92); and Oberholtzer Lee, "Tasting" for more on gustatory taste. The 'travel' of Japanese (and other East Asian) food and cuisine to Europe is analyzed in various articles in Irmela Hijiya-Kirschnereit, ed., *Jahrbuch für Kulinaristik. The German Journal of Food Studies and Hospitality. Wissenschaft – Kultur – Praxis*, vol. 2 (München: Iudicium, 2018).

27 Katarzyna J. Cwiertka, *Modern Japanese Cuisine: Food, Power and National Identity* (London: Reaktion Books, 2006).

28 Lisa Heldke, *Exotic Appetites: Ruminations of a Food Adventurer* (London: Routledge, 2003), p. 49.

29 Marius B. Jansen, ed., *The 19th Century. The Cambridge History of Japan*, vol. 5 (Cambridge: Cambridge University Press, 2007).

30 With Heldke, I would like to define "attitudes to be individual embodiments of culture-wide ideologies" (*Exotic Appetites*, p. 5). Elliott explores in his article the "gap between narrative or representational authority and actual authority in British travel writing" on Japan ("British Travel Writing", p. 198).

31 As such, it contains and conveys ideas of a miniature world (Susan Stewart, *On Longing: Narratives of the Miniature, the Gigantic, the Souvenir, the Collection* (Durham: Duke University Press, 1993)). For the Japanese concept of *washoku*, which in 2013 was added to the UNESCO Intangible Cultural Heritage list and is considered to be a Japanese food culture, providing "the Japanese people with a sense of identity and belonging", see Tine Walravens and Andreas Niehaus, "Introduction: Reconsidering Japanese Food", in Tine Walravens and Andreas Niehaus, eds., *Feeding Japan. The Cultural and Political Issues of Dependency and Risk* (Cham: Palgrave Macmillan, 2017), pp. 1–16 (at p. 1).

32 Bell and Valentine, *Consuming Geographies*, p. 167.

33 Eric Hobsbawm and Terence Ranger, eds., *The Invention of Tradition* (Cambridge: Cambridge University Press, 1983).

34 Bob Ashley et al., *Food and Cultural Studies* (London: Routledge, 2004), p. 86. See also Counihan, and Van Estrik, *Food and Culture*, pp. 289–437.

35 Cwiertka (*Modern Japanese Cuisine*) elaborates that what nowadays is perceived as Japanese cuisine is an invention of the late nineteenth and early twentieth centuries, and it had been and still is an important part of the nation-building process in Japan.

36 Bell and Valentine, *Consuming Geographies*, p. 169.

37 Schaffers, *Konstruktionen*.

38 Arthur Neustadt, *Japanische Reisebriefe. Berichte über eine Fahrt durch Japan* (Berlin: Paul Cassirer, 1913).

39 "Sonderbarerweise war er gar nicht auf die immerhin primitiven Gasthöfe des Inlandes vorbereitet, hatte nichts zu essen bei sich und bemühte sich, gerade als wir eintraten, japanisches Essen hinunterzuwürgen, wobei er eine verzweifelte Anstrengung mit den kleinen Holzstäbchen machte, die ihm Messer und Gabel ersetzen mussten." (Neustadt, *Reisebriefe*, p. 129). Unless otherwise indicated translations of German works are done by myself.

40 During 'diplomatic dinners', which often enough serve as confidence-building measures, the 'Food Question' can be a rewarding topic for small talk in order to circumnavigate other, more precarious topics. What is more, food intake has positive bodily effects, like relaxation, a languorous fatigue, that might help in difficult intercultural and political encounters.

41 Cwiertka, *Modern Japanese Cuisine*, p. 18.

42 Marcel Mauss, "Die Techniken des Körpers", in *Soziologie und Anthropologie* 2, ed. Wolf Lepenies and Henning Ritter (München: Hanser, 1975), pp. 199–220.

43 Georg Simmel, "Soziologie der Mahlzeit" [1910], in Rüdiger Kramme and Angela Rammstedt, eds., *Aufsätze und Abhandlungen 1909–1918*, vol. 1 (Frankfurt am Main: Suhrkamp, 2001), pp. 140–47.

44 The Japanese author Mori Ōgai, who studied medicine in Germany from 1884 to 1888, even performed "nutritional experiments", but, as he wrote, the outcome was not worth recording: "23.–25. Februar / Ich habe Ernährungsexperimente an meinem eigenen Körper durchgeführt [...] Weil die Ergebnisse aber nur unvollkommen sind, werde ich sie nicht veröffentlichen." (*Deutschlandtagebuch 1884–1888*, ed. and trans. Heike Schöche (Tübingen: Konkursbuch, 1992), pp. 19–20).

45 William G. Beasley, *Japan Encounters the Barbarians: Japanese Travellers in America and Europe* (New Haven, London: Yale University Press, 1995), p. 63.

46 Norimasa Muragaki (1813–80), quoted in Beasley, *Japan Encounters*, p. 63.

47 Beasley, *Japan Encounters*, pp. 76–77.

48 Ahmed, *Strange Encounters*, p. 21.

49 A similar conflict (but in a different constellation), which also reveals the complex relation between eating and constructing the 'wild/savage' and the 'civilized man', is described by Oberholtzer Lee, "Tasting", p. 244.

50 Beasley, *Japan Encounters*, p. 77.

51 Cwiertka, *Modern Japanese Cuisine*, p. 24; see also pp. 13–56.

52 During World War I, Brand's meat lozenges were advertised in the *Illustrated London News* as "A Meal in the Vest Pocket" and "Sustaining & Invigorating. A most Acceptable Gift to Officers and Men" (29 May 1915, p. 711).

53 Chamberlain, *Things Japanese*, p. 181.

54 See, for example, Jack Goody, *Cooking, Cuisine and Class: A Study in Comparative Society* (New York: Cambridge University Press, 1982).

55 Cwiertka, *Modern Japanese Cuisine*, p. 33.

56 Christine Ott, *Identität geht durch den Magen. Mythen der Esskultur* (Frankfurt am Main: Fischer, 2017), p. 291–373.

57 Schaffers, *Konstruktionen*, p. 149.

58 Alfred Opitz, *Reiseschreiber. Variationen einer literarischen Figur der Moderne vom 18. bis 20. Jahrhundert* (Trier: WVT, 1997).

59 "der winzige Wirt"; "zehn winzige Dienerinnen" (Max Dauthendey, *Die geflügelte Erde. Ein Lied der Liebe und der Wunder um sieben Meere* (München: Albert Langen, 1910), p. 361).

60 "Wie Kätzlein sich schmiegend, [...] alle schauen weise und altklug auf mich plumpes Europaschwein, / Das mit Stiefeln das Zimmer betritt, statt mit weißseidener Strümpfe Schimmer; das, statt mit Elfenbeinstäbchen zu essen, mächtig mit Messern und Gabeln Fleischhaufen zerschnitt." (Dauthendey, *Die geflügelte Erde*, p. 361).

61 Bird refers to Japanese men as "mannikins" [sic], for example: "At the top of the landing-steps there was a portable restaurant, a neat and most compact thing, with charcoal stove, cooking and eating utensils complete; but it looked as if

it were made by and for dolls, and the mannikin who kept it was not five feet high." (*Unbeaten Tracks*, p. 4).

62 Schaffers, *Konstruktionen*, pp. 58–66.

63 "Es sind die ätherischen Süßigkeiten, die nur den Gaumen eines Embryos im Mutterleib berauschen können, aber unseren Zungen sind sie wie filtrierte und destillierte Luft [...]" (Max Dauthendey, *Mich ruft Dein Bild. Briefe an seine Frau* (München: Albert Langen, 1930), p. 149).

64 Dauthendey, *Die geflügelte Erde*, pp. 381–84.

65 Basil Hall Chamberlain and W. B. Mason, *A Handbook for Travellers in Japan* (London: John Murray, 1891), p. 10.

66 Abbots and Lavis, "Introduction", p. 1.

67 Bernhard Kellermann, *A Walk in Japan. The 1910 Travelogue of Bernhard Kellermann*, trans. Robert Blasiak (Paraparaumu: Fine Line Press, 2015), Kindle Edition (at loc. 1856).

68 "Der Mundraum bildet mithin die Kontaktgrenze von Körperinnenwelt und objekthafter Körperaußenwelt. [...] Weltstoffe müssen durch den Mund ins Innere, [...]. Damit beginnt der auch gar nicht zu überschätzende Vorgang der Verinnerlichung, durch die das Fremde, sofern es 'mundet', in Eigenes verwandelt und, sofern es fremd bleibt, wieder ausgeschieden wird. [...] in der physiologischen Nutrition beginnt die Politik der Assimilation und Dissimilation, der Inklusion und Exklusion." (Hartmut Böhne and Beate Slominski, "Einführung in die Mundhöhle", in Hartmut Böhme and Beate Slominski, eds., *Das Orale. Die Mundhöhle in Kulturgeschichte und Zahnmedizin* (München: Fink, 2013), pp. 11–31 (at pp. 14; 16)).

69 Chamberlain, *Things Japanese*, p. 179.

70 In this context, the very thought of incorporating what is most similar to us, another human being, evokes the strongest repulsion. Anthropophagy is one of the most important food- and eating-related topics in travel writing and within the dynamics and discourses of proximity and distance, inclusion and exclusion, of Othering. It serves as an "ideological device" to "establish difference and construct racial boundaries." (Maggie Kilgour, *From Communion to Cannibalism: An Anatomy of Metaphors of Incorporation* (Princeton: Princeton University Press, 1990), pp. 239–40). The phrase *you are what you eat* includes in that sense two possible perspectives: if eating transforms what is eaten, and it becomes the eater's body, one could ask what is transformed into what by eating another human being. In a more physiological perspective: the other body into mine? Or in a more anthropological, political, or moral perspective: am I becoming the Other while eating? Cannibalism is not a (distinctive) topic in travel writing on Japan, though. What we can observe are metonymical discourses or metaphors that might serve as *cues* to engender overall images that strengthen the process of Othering (see, for example, Dauthendey, *Die geflügelte Erde*, p. 321) without crossing the line to a 'cannibalistic denunciation'.

71 Winfried Menninghaus, *Disgust: Theory and History of a Strong Sensation* (Albany: SUNY Press, 2003), p. 1.

72 Bird's disease (she suffered from back pain from a tumor on her spine) forced her into that unwanted proximity.

73 Paul Rozin and April E. Fallon define disgust as "revulsion at the prospect of oral incorporation of offensive objects. These objects have contamination properties [...]" ("A Perspective on Disgust", *Psychological Review*, 94, no. 1 (1987): pp. 23–41 (at p. 23)).

74 One could even argue that in the process of reading, these articulations could engender a reaction of disgust as being conveyed to the reader's imagination, without any bodily proximity that could cause a 'contamination' of some sort.

75 Charles Darwin, *The Expression of the Emotions in Man and Animals,* ed. by F. Darwin (London: John Murray, 1904), quoted in Sara Ahmed, *The Cultural Politics of Emotion* (New York: Routledge, 2004), p. 82.

76 Ahmed, *Cultural Politics,* p. 82.

77 Ahmed, *Cultural Politics,* p. 82.

10 Going Undercover? Female Bodies and Clothes under Scrutiny in Travel Literature

Sofie Decock

10.1 Introduction

Studies on travel writing frequently address processes of 'othering' in the ways travel writers represent the people they meet on their journey. In doing so, such analyses take into account descriptions of bodily appearances whenever they seem essential and highly salient in such processes. Kerstin Schlieker, for instance, briefly discusses the Islamic veil as an important symbol of 'otherness' in travel texts by the Swiss travel writer and photojournalist Annemarie Schwarzenbach.[1] Overall, however, scholars tend not to engage in systematic and thorough discussions of the role which descriptions of "bodily practices"[2] and clothing performances play in giving meaning to and constructing encounters between locals and travelers. This is somewhat surprising, given that "interaction is first and foremost an interaction of signifying bodies. Through our bodies we perform, express, and (re)present ourselves, and others judge our appearances and performances",[3] as Phillip Vannini and Aaron McCright point out. Indeed, human face-to-face interaction is essentially multimodal, meaning that "various resources [are] mobilized by participants for organizing their action – such as gesture, gaze, facial expression, body postures, body movements, and also prosody, lexis and grammar"[4]. To account for the multimodality of social interaction, the attention has shifted in linguistics and beyond from studying language to studying language and its embodiment, a development also known as the "embodied turn".[5] In an embodied view of language, its verbal dimension is repositioned as one among other modalities, thus overcoming "a logo-centric vision of communication".[6]

Arguably, studying communication from a multimodal perspective is even more relevant in the context of intercultural encounters: as such encounters are often impaired by the lack of a shared tongue, the para- and non-verbal dimensions of communication (prosody, gestures, gaze, facial expressions, appearances) become even more salient. Travel writing typically consists of a series of intercultural encounters, characterized by scrutiny and (potentially ill-informed) interpretation of the other party's bodily practices and appearances as well as by a considerable degree of

DOI: 10.4324/9781003331803-13

self-awareness and self-reflection on the part of travel writers with regard to their own bodily practices and clothing performances and what these might index in terms of social roles and identities.

In this chapter, I intend to examine the social semiotics of bodily practices and clothing performances in the constructions of intercultural encounters through travel writing. In doing so, I choose to focus on the travel texts on Afghanistan written by the Swiss travel writers Ella Maillart (1903–97) and Annemarie Schwarzenbach (1908–42). They traveled together to Afghanistan in 1939, which resulted in Maillart's book *The Cruel Way* (1947)[7] and in several journalistic contributions published in Swiss newspapers by Schwarzenbach. These travel texts were selected for two reasons: first of all, the travel narrators have – as documented in these texts – no command of the local languages. Their inability to speak the locals' tongue not only limits their contacts with them, although these constraints are not (always) explicitly addressed,[8] but it also makes interaction more dependent on para- and non-verbal cues. Secondly, Maillart and Schwarzenbach are both female travel writers, which is believed, from a (post)colonial perspective, to influence interactional power dynamics in complex ways. As Marguerite Helmers and Tilar Mazzeo assert, European women travelers are in the role of both surveyor and surveyed, of seeing subject and seen object.[9] This dual role puts them in an ambiguous position and has the potential of creating intra- and intersubjective tensions related to the transgression of social norms. Based on these selection criteria, their texts can be expected to provide valuable insights into the importance of bodily practices and clothing performances in (the depiction of) intercultural encounters, that is, the signifying processes through which these practices index social roles and a sense of either solidarity or estrangement, how they trigger reflections on travel writers' own embodied performance, how they can reveal power-related tensions and sensitivities, and how they can serve as the medium for confirming, subverting, as well as transgressing social norms.

Of course, studying the multimodality of intercultural encounters in travel writing automatically entails that researchers have to rely on the constructions of these encounters through the perspective of one single participant, the narrator. Indeed, all bodily practices and clothing performances described are conveyed as they are witnessed, experienced, interpreted, remembered, and reconstructed by this narrator, whose writing is moreover influenced by genre-specific expectations. As a consequence, the perspectives of other participants in the encounter are only accessible if these participants make them explicit and the travel writer chooses to (adequately) convey them. This chapter is thus less about studying the fine-grained ongoing process of how multimodal resources are used and create meaning in social interaction in intercultural settings, than about examining the (conventionalized) meanings ascribed to bodily practices and clothing performances in travel writing and the effects of these assumptions and interpretations on how interactional dynamics in these settings are experienced, remembered, and reconstructed in the writing process. Relying on the input given by the narrator also means that only those embodied practices that are topicalized can be addressed in the analysis. In the texts by Maillart and Schwarzenbach on Afghanistan, the focus lies on visual rather than olfactory or haptic stimuli, revealing a "visuo-centric vision of embodiment"[10] on the part of their narrators. As will become clear throughout this paper, the visual stimuli that receive most narrative attention in their work are gestures, facial expressions, and – most notably – clothes.

After introducing the social semiotics of body and clothing as well as the concept of indexicality as theoretical approaches guiding the analysis, I will perform close readings of intercultural encounters as they are depicted in Maillart and Schwarzenbach's texts on Afghanistan. Based on the distinction between European women travelers as surveyed or surveyor and the power asymmetries inherent in this tension (and focusing mainly on the semiotics of female bodies and clothes), the analysis is divided into two parts: a first part on (European) women under the gaze of (Afghan) men and a second part on Afghan women under the gaze of European women.

10.2   Body and Clothes as Semiotic Resources

Following the multimodal character of human interaction, not only language but also the body – its postures, movements, gestures, eye gaze, facial expressions, etc. – as well as the clothes people wear can be considered semiotic resources which people draw on to communicate in their own right. I choose to integrate the social semiotics of the body and of clothing, as both are intimately intertwined in the creation and communication of meaning.[11] Indeed, with clothes physically touching, (partly) covering, and protecting human bodies, they constitute an important part in constructing, regulating, and managing our public physical appearance and the ways our bodily appearance is perceived and judged by others.[12] Together, bodily practices and clothing performances index a certain kind of style, with style being defined "as a multimodal cluster of semiotic practices for the display of identities in interaction".[13]

In developing a theoretical framework geared towards a better understanding of the role of the (clothed) body in semiotic processes, it proves valuable to connect social semiotics, primarily concerned with how meaning arises out of social interaction,[14] with indexicality as conceptualized by Elinor Ochs.[15] Both approaches offer insights into how semiotic resources such as gestures, facial expressions, and clothes come to be associated with social categories, stressing that such associations, although they can be subject to conventionalization and stereotyping, are never fixed but can change over time. In averting deterministic views on the relation between signs and their meanings, Ochs proposes a direct and indirect level of indexicality.[16] On a direct level, an embodied movement such as lowering the head might index a modest stance, whereas on an indirect level, lowering the head becomes ideologically associated with particular social groups believed to take such modest stances (for example, women in patriarchal societies). Although over time, such associations may come to be perceived as direct, it is important not to lose sight of these two levels of indexicality: indeed, semiotic resources index social categories *indirectly*, not directly. Moreover, as Mary Bucholtz points out, semiotic resources "are generally associated not with broad social categories like women but rather with more specific social types and personas",[17] such as 'Islamic women from Afghanistan'.

The interplay between semiosis and gender ideologies already indicates how "power plays a crucial role in the process of assigning meaning to signs".[18] Indeed, semiosis as it takes place in social interaction is hugely shaped by power dynamics. There are not only power imbalances in relation to the cultural value which is ascribed to specific signs, but also in relation to the social norms pertaining to the use of semiotic resources when signifying bodies interact. The direction of the gaze, for instance, is ideologically regulated, as Patrizia Calefato reminds us, as "'to look at' and 'to be looked at' are conditions with uneven hierarchical positions: the former being proper to men, the latter to women".[19] The same goes for the relation between Western and non-Western cultures: "the western world looks at, and the 'non-western' is looked at".[20] In the case of European women traveling to non-Western countries in the twentieth century, this results in the ambiguous position of being in the role of both surveyor and surveyed, as indicated above. Moreover, power dynamics are made even more complex by the fact that all 'signifying bodies', no matter what their assigned place in the power hierarchy is, are at the same time individual agents who can deploy semiotic resources for their own interactional purposes; they may for instance deploy them in such a way as to exploit or set aside conventionalized associations.[21] When it comes to clothes in particular, they are known for being used to support gender and class ideologies and to mark clear gendered and social dichotomies.[22] Especially in collectivist societies, clothes are narrowly regimented semiotic resources. They are mainly used to communicate messages about the gender, class, religion, and ethnicity of the wearer, and less for individual self-expression. Overall, it can be maintained that, while keeping in mind the direct and indirect levels of indexicality and the potential for change, clothed bodies are used to perform and express various kinds of social roles and identities, and these appearances and performances are judged and interpreted by observers.

10.3   Travel Writing of Annemarie Schwarzenbach and Ella Maillart on Afghanistan

Before analyzing Schwarzenbach's and Maillart's texts on Afghanistan from the perspective of body and clothing semiotics, I will first introduce both travel writers. Ella Maillart is considered an important twentieth-century traveler, travel writer, and photographer. Her travels took her to Moscow, the Caucasus, the former Soviet Republics of Central Asia, China, India, Afghanistan, Iran, Nepal, and Turkey. She financed her trips by working as a writer and journalist. Her first travel report, *Parmi la Jeunesse russe*, appeared in 1932,[23] preceding many others. Annemarie Schwarzenbach was known to her contemporary readership, above all, as a travel and photojournalist, but she also wrote short stories, novels, and even poetry. After the (re-)discovery of her literary work in 1987, one can now safely claim that she ranks within the canon of German-speaking female writers. In the 1930s and 1940s, and before her sudden death in 1942, she traveled through Europe, Asia, the Americas, and Africa. Although she received financial support from her extremely wealthy Swiss family, she also took on assignments to compose articles and features. Her journalistic writings appeared in the most prominent Swiss–German newspapers and magazines of the time. In 1939, Maillart and Schwarzenbach embarked upon a journey to Afghanistan together. After thoroughly preparing for their trip, they left on 6 June 1939 from Geneva in a Ford car and arrived in Kabul after a 12-week journey in early August of the same year. In Afghanistan itself, they had traversed the steppe and deserts of Afghanistan–Turkestan and the Hindu Kush, visiting places such as Herat, Mazari Sharif, Puli Khumri, the Valley of Bamyan, and Begram. In terms of publications and financial benefits, the journey turned out to be a disappointment for Ella Maillart because only a few of her articles were published in Switzerland. Annemarie Schwarzenbach, on the contrary, was able to publish more than 50 features, reports, and photo stories. Among others, her articles were published in the *National-Zeitung*, *Weltwoche*, and *Neue Zürcher Zeitung*. The photo stories were published in French or in German, mostly together with Ella Maillart, in *Sie und Er* and in *Zürcher Illustrierte*. A selection of these articles can also be found in the compilations *Auf der Schattenseite*,[24] in *Alle Wege sind offen*,[25] and in *Orientreisen*.[26] Apart from journalistic articles, Annemarie Schwarzenbach also wrote the narrative volume *Die vierzig Säulen der Erinnerung*,[27] inspired by this trip, and Ella Maillart wrote her travelogue *The Cruel Way*,[28] which came "to be seen as a classic work of travel literature relating to Afghanistan".[29]

It is common knowledge that travel literature is a hybrid and heterogeneous genre. Monika Fischer characterizes Ella Maillart and Annemarie Schwarzenbach's travel writing as "a cross between literary essay, photojournalism, travel accounts, testimony and personal soul searching.

Schwarzenbach, in particular, presents an interesting example of liminal writing that is lyrical and poetic".[30] Indeed, Schwarzenbach's travel writing is known for its tendency to transgress the frames of fact-based writing and to explore the spaces of experienced reality and imagination in its depiction of an 'inner' journey. In that sense, her travel texts often break with generic conventions of the travel reportage and the travel report: travel journalists are mainly expected to impart knowledge and facts to their readers while at the same time entertaining them with interesting, well-written 'stories'. This does not mean, however, that subjectivity and introspection dominate all of her writings. Some texts are rich in historical facts and knowledge of the current political circumstances, demonstrating that as a writer, she was able to slip into different roles and adapt each role to the expectations of her target readership. Compared to Schwarzenbach, Maillart's travel writing is much less introspective and almost entirely fact-based, descriptive, stylistically objective, commentarial, or – as Schlieker classifies it – 'ethnographic'.[31]

Interestingly, when it comes to the texts produced by the two women based on their journey to Afghanistan, it can be observed that they influenced each other in their writing. It is quite clear that Schwarzenbach focuses more on the local population than in her previous travel texts, and Maillart's *The Cruel Way*, although it can still be considered a travelogue through a chronological depiction of the journey interspersed with historical and sociocultural background information, reflects more on the inner life of the narrator and reports less on encounters with locals.

As previous studies have amply shown, the texts both writers produced against the backdrop of their travels together participate in but also co-construct colonialist, orientalist, imperialistic, and racist discourses associated with the 'Orient' from a Western perspective,[32] a discursive tradition to which European women – in spite of their own history of oppression – are indeed in no way immune to. Several orientalist topoi are featured in Maillart's and Schwarzenbach's writings on Afghanistan. Most notably, their travel writers imagine a pre-capitalist, even pre-civilizational community of 'free people' living in quasi-paradisiacal, isolated areas, an idealization which goes hand in hand with their pessimistic assessment that Europe has been hopelessly spoilt by capitalism, industrialization, and war, and that the Orient is awaiting a similar faith as it is already being confronted with the first signs and destructive consequences of industrialization. At the same time, and paradoxically, their travel writers also evoke typical negative images of the Orient, as reflected in relatively frequent statements about the oppressed state of women and the lack of education, phenomena which are considered to be at odds with Western cultural values.[33]

In addition to exploring orientalist discourses in Maillart's and Schwarzenbach's texts on Afghanistan, previous studies have already focused on the depiction of women from a Critical Discourse Analysis (CDA) perspective[34] and on the topicalization of language and verbal communication in the construction of intercultural encounters in their texts.[35] With regard to the latter research focus, it was observed that their travel narrators have no command of the local languages, a circumstance which made para- and non-verbal cues more salient in their interactions with locals. Against this background, it proves all the more interesting to systematically examine the role of bodily practices and clothing performances in the construction of intercultural encounters in their texts. With this goal in mind, I carefully read both writers' texts on Afghanistan and selected a corpus of relevant text excerpts, that is excerpts in which bodily practices and clothing performances are described against the background of intercultural contact and experience. For Maillart, all excerpts stem from her travelogue *The Cruel Way*, whereas for Schwarzenbach, the excerpts are taken from her travel reports "Two Women alone in Afghanistan" (*All the Roads* 89–97), "In the Garden of the Beautiful Girls of Qaisar" (*All the Roads* 51–55), "The Women of Kabul" (*All the Roads* 56–60), and "The Neighbouring Village" (*All the Roads* 63–69).[36]

10.4   Part I: (European) Women under the Gaze of (Afghan) Men

In the first half of the twentieth century, women traveling alone were still seen as transgressing the norm from a European perspective and even more so from the perspective of people living in Afghanistan, a country entrenched in tribal patriarchy,[37] since it was believed that women are supposed to be tied to their home, and that their bodies and minds are not fit for excessive movement.[38] Female travelers often refer to this norm directly or indirectly in order to legitimate their choice to travel. In Schwarzenbach's travel report "Two Women alone in Afghanistan", the narrator seems to allude to this norm when she concedes how many friends and acquaintances had asked them about how it had been possible for them to undertake the endeavor of traveling alone as women in a country like Afghanistan. She uses this as an opportunity to claim that they had experienced nothing but hospitality, with the exception of one incident in which their camera was stolen but which concluded with a happy ending. In all of Schwarzenbach's texts on Afghanistan, no problems they experienced while traveling on the basis of their sex or their appearance are mentioned. This version of events does not entirely match the narrator's experience in Maillart's *The Cruel Way*, who does report on the strategies they consciously employed to avoid or avert trouble. Her narrator refers indirectly to the norm-transgressive dimension of their trip when she writes:

> Another kind of foresight had made me persuade Christina to buy a skirt: I had convinced her that as long as she wore her grey bags she

would be taken for a man and Afghan harems would remain closed to her. I was also convinced that when difficulties are encountered in Asia, women are more readily helped if they are seen to be without a man.

(Maillart, *The Cruel Way* 98)[39]

The quote reveals that 'Christina' was someone who did not usually wear skirts, and it is implied in this instance that the habit of wearing skirts unambiguously indexes femininity. Maillart's travel narrator gives her travel companion the advice to perform a clear feminine gender identity, indexed by the skirt, as she believes this would bring considerable advantages in terms of access and support in the strictly patriarchal context they find themselves in. The advantages she names, access and support, echo previous statements of female travel writers such as Cristina Trivulzio and Ida Pfeiffer.[40] Her recommendation can be read as an implicit legitimation of travel in that the travel narrator implicitly turns around the common argument that traveling would be too dangerous for women. Ironically, this legitimation relies on the same patriarchal premise as a 'travel ban' for women, namely their supposed weakness: because women are considered to be weak, they receive more help when traveling, which makes their travels safer, and which in turn makes it perfectly fine for them to travel.

Judging from the examples of intercultural encounters in Afghanistan which Maillart's narrator describes, her foresight indeed seems to have been quite accurate. Not only are they given access to the house and garden of the mayor of a village, where his wife and daughters live, but she also describes several instances of how they enjoy or are offered help, support, and hospitality by Afghan men. The desire of Afghan men to protect them because they are women traveling alone is mentioned explicitly in one case: "When we took leave, White-Beard offered to accompany us to Kabul: women should not travel alone!" (Maillart, *The Cruel Way* 177). Moreover, their first encounter with Afghan men seems to serve as a confirmation that bodies which do not allow for unambiguous gender ascriptions are a recipe for trouble in this particular environment. Indeed, upon crossing the border between Iran and Afghanistan at night, the following happens:

[T]wo men [...] were pointing their rifles at us. One soldier [...] crouched between our mudguard and bonnet. The other [...] dropped plump between us [...]. I was distracted by our new neighbour and by what must have been going on in his mind. He could not make out what these gibbering people were. [...] Could they possibly be female, the two of them? The soft reflection coming from the dials on the switch-board lit only the lower part of our small chins which

> looked feminine. But there was short hair on both heads. ... Attempts
> at conversation having yielded no elucidation, the Afghan decided to
> solve the mystery according to his means: very slowly his hands began
> to follow the curves of our ribs! Had there ever been a more ridiculous
> situation? We could not afford to be offended and thus perhaps antag-
> onise these feline soldiers because after all, we were at their mercy.
>
> (117–18)

The narrator interprets the behavior of the Afghan soldiers who cross borders of bodily intimacy as rooted in confusion, because, given the nightly
and spatial circumstances (leaving only certain parts of the body visible)
and their unfamiliarity with European women, their female features must
have been difficult to discern. In an assumed search for signs of female or
male anatomy, the soldiers start touching them, a transgressive act which
they feel obliged to tolerate.

Although Maillart's travel narrator claims that being clearly identifiable
as female comes with certain privileges when traveling in Asia, it turns
out not to be an all-round advantage as it also simultaneously *subjects*
them to the male gaze. On two occasions, their presence at all-male public
gatherings seems compromised. The following scene, in which they watch
a duel between two stalwarts on their way back from the bazaar, depicts
how their access to such sites is constrained: "We walked away: we were
attracting too much attention and, for my part, I wished my skirt were
longer. Overtaking us, the manager of the rest-house summed up the situation: 'Remember that these men have so far seen the faces of only four
khanums – their mother, sister, wife and daughter'" (128). The narrator
is rather vague in describing this incident and it remains unclear how they
sensed that they were attracting too much attention and whether they were
forced to leave or not. Stating, however, that she wished her skirt were
longer can be interpreted as an implicit admission that she felt uneasy at
being subjected to the male gaze, a gaze which indexes the message that she
is transgressing the patriarchal norm of female modesty by being dressed
as she is. Interestingly, it is again the skirt that marks femininity, but in
this case, it is associated with a disadvantage rather than an advantage, a
shift which is linked to the different role she assumes in this scene: not the
role of a woman traveler in need of help, but of a woman traveler merely
observing practices connoted with masculinity and thus being perceived as
an intruder violating the patriarchal norm of gender segregation.

Another awkward and uncomfortable situation occurs when the two
women are guests of the governor of an Afghan province. This situation
indicates that performing an unambiguous female identity does increase
hospitality, but it comes at the prize of potentially being turned into an
object of sexual desire by their Afghan hosts. In this case, the advances

are still made in a respectful and therefore not terrifying way, so that the narrator – at least in the way she chooses to render the situation – is able to take it with humor:

> Before dawn we heard quite near us the voice of our host calling 'Madame!' [...] Tacitly, we postponed answering him till the sun rose. [...] The governor apologised for waking us. [...] Then, with charm and simplicity, looking beyond both of us, he said: 'I would have liked to breathe the flowery fragrance of your face!' I decided that the remark was not for me, my face having been so far likened only to horse, ship's bow or wild grass.
>
> (138–39)

Up to now, the narrator in Maillart's *The Cruel Way* refers to their traveling bodies as wearing a skirt when it comes to positioning themselves as female travelers in the eyes of Afghan men. This changes, however, when the narrator finds herself in the garden of the mayor where they can meet his wife and daughters. In her conversation with these Afghan women, the narrator refers to her own body and that of her travel companion Christina as wearing trousers, as the following quote illustrates:

> [T]he mayor of the village [...] wanted us to share the meal of his womenfolk. We were soon sitting on a carpet under a mulberry-tree, surrounded by charming people. Old fashion-papers appeared, our opinion was wanted and, after some shy nudgings, a length of blue silk was brought which we were asked to cut out for the young daughter of the house. Our mimicry explained that since we wore trousers we knew nothing about frocks.
>
> (143)

This contrast between skirts and trousers indicates that, as Monica Owyong states, clothes "have always been used to mark the dichotomy between the powerful and the weak".[41] While Maillart's reference to wearing skirts connotes weakness, in this case embodied by women, her reference to wearing trousers indicates a power position, in this case embodied by Europeans. Of course, the fact that trousers are associated with power, even if women have come to wear them, still shows the power imbalance between men and women, or to put it in Owyong's words: "That women have adopted what was characteristically male clothing [...] while men have not sought with similar vigour to adopt the female dress or skirt reinforces the power discrepancies between men and women."[42]

Overall, the narrator's depiction of her clothing practices – and to a much lesser extent her bodily practices – reveals context-dependent strategies, that is strategies depending on whom she is interacting with and the role she assumes in a certain situation. While she accommodates patriarchal norms by leaving all-male sites in which she feels unwanted based on her sex, she also resists these norms when declaring to her female Afghan audience that she usually wears trousers, thus indexing the identity of a liberated European woman. In addition, she also combines accommodation and resistance in a paradoxical way: on the one hand, she advocates the performance of a clear female identity which conforms to societal norms in terms of dress code (that is, wearing a skirt as a woman), but on the other hand, she frames this as a strategy which legitimates her norm-transgressive endeavor of traveling alone as a woman towards her European readership, and which brings privileges while traveling in Afghanistan in terms of support and access to all-female sites. These context-dependent strategies can be conceptualized as "strategic ambiguity"[43] on the part of the female travel writer. Strategic ambiguity as an agency-related concept goes beyond the "accommodation/ resistance paradigm" in that it unites "rebellious and submissive life-choices under one concept, making possible the give-and-take that actually happens in one's life".[44] The concept allows one to discuss how women in patriarchal settings maneuver between accommodation and resistance, "creating a space in which they make varied choices and express varied perspectives".[45]

10.5   Part II: Afghan Women under the Gaze of European Women

While reflections on the strategic performance of their traveling bodies are only made by the narrator in Maillart's *The Cruel Way*, descriptions of Afghan women can be found in both Schwarzenbach's and Maillart's writings on Afghanistan. The topic of women in Afghanistan is more prominent in Schwarzenbach's work, with three travel reports dedicated to it: "In the Garden of the Beautiful Girls of Qaisar", "The Women of Kabul", and "Der Tschador".[46] In their texts, the narrators admit to not seeing many women during their trip, and very few encounters with them are described ("But we seemed to be in a land without women!", Schwarzenbach, *All the Roads* 51; "you've come to the country where women are not seen", Maillart, *The Cruel Way* 137). When they do describe them, the chador, which is nowadays commonly known as the burka,[47] serves – not surprisingly – as an important piece of clothing to distinguish themselves from Afghan women, and to distinguish nomadic Afghan women from non-nomadic Afghan women. From these descriptions, it becomes clear that the non-nomadic Afghan women wearing the chador in public trigger a deep sense of estrangement and repulsion in the narrators:

> We knew the chador, Mohammedan women's all-enveloping drapery, which has little in common with romantic notions of oriental princesses' wispy veils. It hugs the head, perforated at the face in a sort

of grille, then falls to the ground in voluminous folds, barely revealing the embroidered tips and worn-down heels of the slippers. We
saw these muffled, formless figures darting shyly down the lanes of
the bazaar and knew they were the wives of the proud, free-striding
Afghans with their love of company and jovial conversation who spent
half the day lounging in the teahouse and at the bazaar. But there was
little humanity in these ghostly apparitions. Were they girls, mothers,
crones, were they young or old, happy or sad, beautiful or ugly? How
did they live, what occupied them, who received their sympathy, their
love or their hate?

(Schwarzenbach, *All the Roads* 51)

Now and then we passed a few of these hidden women-shrouded sil-
houettes guiding their steps from behind peepholes embroidered lat-
tice-like before the eyes. Driving, we found them a public danger: they
saw little and heard even less. We had to be right upon them before
they would jump aside, frightened like crackling hens.

(Maillart, *The Cruel Way* 128–29)

The process of 'othering' at work in these two quotes is extreme, up to the
point that these Afghan women wearing the chador in public lose their
individuality and are dehumanized. Quite tellingly, the narrator in *The
Cruel Way* compares the chador to a shroud, which is a piece of tissue put
on someone who died, evoking the image of women who are muted and
almost dead. In itself, the chador as a piece of clothing indeed does not
invite human interaction, as it covers the face and the rest of the body in
their entirety, restricting the use of semiotic resources which people draw
on to communicate. In that sense, the cultural imperative of having to
wear it as a woman in public clearly constitutes a disciplinary act in that
it severely reduces women's agency in taking part in public social life. The
chador could be conceptualized as a uniform because it functions as a
regulatory apparatus, largely impersonalizing individual women's bodies
and disciplining them into a hierarchical social order.[48] Apart from the
materiality and uniform-like quality of the chador and its communicative
implications, these descriptions of veiled Islamic women are part of a long-
standing tradition of negative Western reactions to the Muslim veil. In this
tradition, the veil has come to stand for a religious corporeality as opposed
to the secular or enlightened corporeality of the West, and as a religious
symbol it is looked upon with suspicion and perceived as an alarming signal that interrupts the automatic flow of everyday signs.[49]

Influenced by this Western topos of the Muslim veil as a disturbing marker of oppression, European women wearing a chador in
Schwarzenbach's and Maillart's texts attract the narrators' attention. In

Schwarzenbach's text "The Women of Kabul", the narrator experiences "a crippling sense of horror, almost aversion" (*All the Roads* 56) when meeting such a woman. In her depiction of the woman, she frames her as an oppressed being who has lost all sense of taste, individuality, personality, and self-esteem. She is referred to as "the taciturn guest" who "ate cake without stopping" (58–59), making it seem as though she had been pressed into silence, 'eating her words'.[50] The woman's European clothes underneath her chador are judged by the narrator as "not very tasteful[l] and somewhat slopp[y]" (Schwarzenbach, *All the Roads* 58–59). Her face is described as "strangely expressionless" to the extent that it "might as well have been hidden behind the veil", and "even when she wept, all that she showed was misery, no real sorrow, no regret, no unbroken defiance" (58–59). In what comes across as a violent interpretive act, the narrator projects well-known orientalist stereotypes on the woman's body, all the more so in describing her face as "oriental" (59) and equating this label with passivity. While the narrator in Schwarzenbach's travel report interprets a European woman wearing a chador as an inconceivable symbol of quasi-absolute submission, the narrator in *The Cruel Way* allows for a more nuanced view in her depiction of such a woman employing 'strategic ambiguity':[51] she is not associated with passiveness as in Schwarzenbach's text but is described as walking "alone, decidedly", as having an "unyielding" face, and as someone who "thought the time was ripe for the chadur to be given up; but it was not for the foreign women married to Afghans to show the way unless they wanted their lives to become impossible" (Maillart, *The Cruel Way* 187), thus revealing both her emancipatory stance and the disciplinary forces which she simultaneously obeys.

As mentioned before, the narrators in *The Cruel Way* and "In the Garden of the Beautiful Girls of Qaisar" both write about the singular opportunity they had to visit non-nomadic Afghan women – a mother and her daughters – in their private environment, at the home of an Afghan mayor. In their accounts, it is particularly the young women's bodies which are imagined as sites of struggle, with their clothing practices marking the dichotomy between oppression and lightness and between tradition and modernity. When at home, these women do not wear a chador anymore, but a veil. Contrary to the way the chador is described, the descriptions of the materiality of these veils evoke associations of lightness: "white wimple", "a softening white veil" (Maillart, *The Cruel Way* 143) and "light, airy veils" (Schwarzenbach, *All the Roads* 52). Moreover, the Afghan girls seem to wear modern female European clothing requiring no further elaboration: "summer dresses" (52), "pink skirt, white blouse" (Maillart, *The Cruel Way* 143). As illustrated in an earlier quote in Section 10.4 (143), modern European clothing also becomes a topic of conversation during this encounter. The young Afghan women would like the European women to cut them a dress in the European fashion. Against the background of a society which imposes strict rules on a woman's garments, it thus seems that these Afghan girls are flirting with a more European, modern lifestyle, at least in their private environment. This norm-transgressive aspiration, though still limited in scope as no other signs of norm-breaking behavior were picked up by the narrators, equally suggests 'strategic ambiguity' on the part of these Afghan girls. Their request is welcomed by both narrators, and most explicitly in the following quote: "And by sending the promised dress patterns from Kabul to Qaisar, we contributed in our own small way towards the consequences of these laws. We fought the chador!" (Schwarzenbach, *All the Roads* 55).[52] It should be noted that when chador-wearing women are presented as applying strategic ambiguity in Maillart's and Schwarzenbach's texts, their acts of resistance tend to be associated with rejecting the veil. In more recent scholarship, such Eurocentric views have been criticized, and it has been shown that, admittedly within the confines of patriarchal societal structures, there also lies emancipatory potential in the act of veiling itself. In this respect, the veil might enable freedom of movement, protect against harassment, and allow for the expression of individual spirituality and/or a political identity.[53]

As described in the introduction to *The Routledge International Handbook of Veiling and Veiling Practices*, veiling is an old practice which exists in many different types and which can be ascribed different meanings, potentially indexing categories such as gender, social class and religion.[54] As illustrated above, it is particularly the chador as a type of veil which tends be associated with negative meanings in Maillart's and Schwarzenbach's texts. This is probably due to the chador's particular materiality, which constrains human multimodal interaction, in combination with Europe's tradition of negatively stereotyping the Muslim veil in general. Other veiling practices, however, receive more positive connotations in their writings. The narrator in Maillart's *The Cruel Way*, for instance, connects the assumed pride of Afghan men to their turbans, as the following quote shows:

[T]his turban-wearing country has much in its favour: since men are influenced by what they wear, the heads of its people are proudly erect. If by winding a length of cloth around his skull the Afghan gains in dignity and beauty, would he not be a fool to adopt the ragged cap of the Persians?

(128)

Whereas the chador is presented as a piece of clothing which pulls down the human body, thus bringing it into submission, the turban is presented as a phallic symbol in that it is a piece of clothing which lifts the body up, thus bringing it into elevation. Highly fascinating with regard to the semiotics of veiling practices is also the depiction of nomadic Afghan women in their texts. These women are explicitly referred to as being unveiled (167; Schwarzenbach, *All the Roads* 66), thus creating a strong discursive contrast with the veiled non-nomadic Afghan women. For both types of women, (un)veiling and bodily practices seem to be in semiotic alignment with each other. While veiled women's bodily practices are associated with deference (see "shy", Maillart, *The Cruel Way* 143; "shyly", Schwarzenbach, *All the Roads* 51), the unveiled women's bodily practices, however, suggest power, as the following quotes indicate: "The women moved easily and with commanding gestures, continually pulling up the great black cloth that fell loosely from the top of the head" (Maillart, *The Cruel Way* 180); "Yes, for the first time in ages I hear the voice of a woman, cheerful and imperious" (Schwarzenbach, *All the Roads* 66). Strikingly, these nomadic women are explicitly categorized as unveiled; however, this does not necessarily mean that they do not participate in veiling practices, as the reference to the head-cloth shows. This indicates that the signifier 'veil' seems to be reserved to non-nomadic Afghan women, while a different signifier ('cloth') is used to describe possible veiling practices of nomadic Afghan women. Interestingly, this does not mean that these nomadic women were not Islamic. They probably were, but the religious identity of these women seems to be considered irrelevant in the light of their nomadic traditional identity. By making this artificial distinction between veiled women (which are primarily identified as Islamic) and unveiled women (which are primarily identified as nomadic) on the semiotic level, two different traditions are evoked and clearly separated from one another: an Islamic vs. an Indo-European tradition in Afghanistan. In line with specific orientalist discourses originating in a German-speaking context, the latter tradition is perceived more positively in the sense of being more authentic and with a closer proximity to European cultural heritage. This cultural proximity is addressed more explicitly in Schwarzenbach's work, in which similarities in social structures and values (such as freedom and democracy) are observed between nomadic communities living in the Afghan mountains and Swiss pastoral traditions.[55]

When comparing the texts of Maillart and Schwarzenbach in their renderings of Afghan women's bodily practices and clothing performances, the discursive similarities are quite apparent, and the narrators also share a limited ethnographic stance, with little interaction with local women being described, and with superficial rather than thick descriptions. At the same time, however, it does seem that orientalist stereotypes about women in Afghanistan are produced more intensely in the travel reports of Annemarie Schwarzenbach, creating the impression that in these texts, local women are mainly being instrumentalized in order to idealize 'the authentic Afghanistan' with its free and nomadic people and its cultural similarities to 'original Swiss values' at the expense of Islamic traditions.

10.6 Conclusion

The aim of this chapter was to systematically examine the social semiotics of bodily practices and clothing performances in the depiction of intercultural encounters in Ella Maillart's and Annemarie Schwarzenbach's travel writing on Afghanistan, and to demonstrate the importance of these practices in intercultural impression management, relationship building, power dynamics and the strategic negotiation of social norms. As the first part of the analysis has shown, the female narrator in Maillart's *The Cruel Way* reflects on and experiences the gendered performance of herself and her travel companion's bodies and clothes (symbolized by the 'skirt') in a foreign environment and how 'going undercover' in their case seems to be neither possible nor desirable. They transgress traditional gender norms by strategically exploiting them to their advantage, but, at the same time, both gender ambiguity and a clear female identity have the potential of bringing them into uncomfortable situations in which their bodies are objectified. The second part of the analysis has focused on various instances of 'othering' in descriptions of local women by the European female narrators in Maillart's and Schwarzenbach's texts. In this power constellation, the female narrators construct themselves as wearing trousers (instead of skirts), while the depiction of Afghan women concentrates on (the absence of) veiling practices and body postures associated with them. Central to this 'othering' is the dichotomy between tradition and modernity, which plays out in different ways, and norm transgressions with regard to this dichotomy. Overall, it is remarkable how the traveling bodies in both women's writings on Afghanistan, and arguably in most if not all travel writing, are found to discursively navigate the opposition between tradition and modernity, an opposition which, in that sense, can be considered a constitutive frame against which traveling bodies position themselves and others.

Notes

1 Kerstin Schlieker, *Frauenreisen in den Orient zu Beginn des 20. Jahrhunderts: weibliche Strategien der Erfahrung und textuellen Vermittlung kultureller Fremde* (Köln: WiKu-Verlag für Wissenschaft und Kultur, 2003), p. 179–80.

2 Mats Eriksson, "Referring as Interaction: On the Interplay between Linguistic and Bodily Practices", *Journal of Pragmatics*, 41, no. 2 (2009): 240–62 (at p. 240).

3 Phillip Vannini and Aaron M. McCright, "To Die for: the Semiotic Seductive Power of the Tanned Body", *Symbolic Interaction*, 27, no. 3 (2004): 309–32 (at p. 312).

4 Lorenza Mondada, "Challenges of Multimodality: Language and the Body in Social Interaction", *Journal of Sociolinguistics*, 20, no. 3 (2016): 336–66 (at p. 338).

5 Mondada, "Challenges of Multimodality", p. 338.

6 Mondada, "Challenges of Multimodality", p. 336.

7 Ella Maillart, *The Cruel Way: Switzerland to Afghanistan in a Ford, 1939. With a New Foreword by Jessica Crispin* [1947] (Chicago: University of Chicago Press, 2013).

8 Sofie Decock and Uta Schaffers, "'… such are the joys of travelling when you do not speak the language of the country you are passing through': Zugänge zur anderen Kultur in Reisetexten von Frauen der 1930er Jahre", in Michaela Holdenried, Alexander Honold, and Stefan Hermes, eds., *Reiseliteratur der Moderne und Postmoderne* (Berlin: Erich Schmidt, 2017), pp. 65–80 (at pp. 76–77).

9 Marguerite Helmers and Tilar Mazzeo, "Introduction: Travel and the Body", *Journal of Narrative Theory*, 35, no. 3 (2005): 267–76 (at p. 269).

10 Mondada, "Challenges of Multimodality", p. 336.

11 Monica Owyong, "Clothing Semiotics and the Social Construction of Power Relations", *Social Semiotics*, 19, no. 2 (2009): 191–211 (at p. 192).

12 Vannini and McCright, "To Die for", p. 313; and Patrizia Calefato, "Fashion as Cultural Translation: Knowledge, Constrictions and Transgressions on/of the Female Body", *Social Semiotics*, 20, no. 4 (2010): 343–55 (at p. 344).

13 Mary Bucholtz, "From Stance to Style: Gender, Interaction, and Indexicality in Mexican Immigrant Youth Slang", in Alexandra Jaffe, ed., *Stance. Sociolinguistic Perspectives* (Oxford: Oxford University Press, 2009), pp. 146–70 (at p. 146).

14 Vannini and McCright, "To Die for", p. 315.

15 Elinor Ochs, "Indexing Gender", in Alessandro Duranti and Charles Goodwin, eds., *Rethinking Context: Language as an Interactive Phenomenon* (Cambridge: Cambridge University Press, 1992), pp. 335–58 (at p. 335).

16 Ochs, "Indexing Gender", pp. 341–43.

17 Bucholtz, "From Stance to Style", p. 148.

18 Vannini and McCright, "To Die for", p. 313.

19 Calefato, "Fashion as Cultural Translation", p. 343.

20 Calefato, "Fashion as Cultural Translation", p. 343.

21 Bucholtz, "From Stance to Style", p. 147.

22 Owyong, "Clothing Semiotics", p. 196; and Calefato, "Fashion as Cultural Translation", p. 344.

23 Ella Maillart, *Parmi la Jeunesse Russe* (Paris: Fasquelle Editeurs, 1932).

24 Annemarie Schwarzenbach, *Auf der Schattenseite. Ausgewählte Reportagen, Feuilletons und Fotografien 1933–1942*, eds. Regina Dieterle and Roger Perret (Basel: Lenos, 1990).

25 Annemarie Schwarzenbach, *Alle Wege sind offen. Die Reise nach Afghanistan 1939/1940*, ed. Roger Perret (Basel: Lenos, 2000); Annemarie Schwarzenbach, *All the Roads Are Open. An Afghan Journey 1939–1940*, trans. Isabel Fargo Cole (Kolkata: Seagull Books, 2011). Further references to this translated edition are included parenthetically in the text.

26 Annemarie Schwarzenbach, *Orientreisen. Reportagen aus der Fremde*, ed. Walter Fähnders (Dortmund: edition ebersbach, 2010).

27 Annemarie Schwarzenbach, *Les Quarante Colonnes du souvenir/Die vierzig Säulen der Erinnerung*, ed. Dominique Laure Miermont (Noville-sur-Mehaigne: Esperluète, 2008).

28 See endnote 7. Further references to this edition are included parenthetically in the text.

29 William Maley, "Afghanistan as a Cultural Crossroads: Lessons from the Writings of Ella Maillart, Annemarie Schwarzenbach and Nancy Hatch Dupree", *Asian Affairs*, 44, no. 2 (2013): 215–30 (at p. 217).

30 Monika Fischer, "Travel Writing and Parrhesia: The Case of Annemarie Schwarzenbach and Ella Maillart", *TRANS. Internet-Zeitschrift für Kulturwissenschaften*, no. 16 (2005) (last accessed May 24, 2022).

31 Schlieker, *Frauenreisen in den Orient*, p. 151.

32 See, for example, Schlieker, *Frauenreisen in den Orient*; Daniela Gretz, "'Wie der Orient sich im Auge einer Tochter des Okzidents abspiegelt'. Frauen-Reisen in den Orient von Ida Pfeiffer bis Ella Maillart", in Martin Tamcke and Arthur Manukyan, eds., *Protestanten im Orient* (Baden-Baden: Ergon, 2009), pp. 165–90; Sofie Decock, *Papierfähnchen auf einer imaginären Weltkarte: mythische Topo- und Tempografien in den Asien- und Afrikaschriften Annemarie Schwarzenbachs* (Bielefeld: Aisthesis, 2010); and Sofie Decock and Uta Schaffers, "Die Darstellung der 'anderen Frau' in Reisetexten", *Zeitschrift für Literaturwissenschaft und Linguistik*, 48, no. 4 (2018): 775–97.

33 Decock and Schaffers, "Die Darstellung der 'anderen Frau' in Reisetexten", pp. 775–97.

34 Decock and Schaffers, "Die Darstellung der 'anderen Frau' in Reisetexten", pp. 775–97.

35 Decock and Schaffers, "… such are the joys of travelling", pp. 65–80.

36 These travel reports were originally published in German in Swiss newspapers: "Zwei Frauen allein in Afghanistan [Two Women alone in Afghanistan]", *Thurgauer Zeitung* (16/17 February 1940); "Das Nachbardorf [The Neighbouring Village]", *National-Zeitung* (18 January 1940): 29; "Die Frauen Afghanistans [In the Garden of the Beautiful Girls of Qaisar]", *National-Zeitung* (13/14 April 1940): 172; "Die Frauen Kabuls [The Women of Kabul]", *National-Zeitung* (27/28 April 1940): 196.

37 Valentine M. Moghadam, "Revolution, Religion, and Gender Politics: Iran and Afghanistan compared", *Journal of Women's History*, 10, no. 4 (1999): 172–95 (at p. 172).

38 Ulrike Stamm, "Zwischen Anpassung und Widerstand. Die Verhandlung von Normbrüchen in Reiseberichten von Frauen", in Uta Schaffers, Stefan Neuhaus, and Hajo Diekmannshenke, eds., *(Off) The Beaten Track?* (Würzburg: Königshausen & Neumann, 2018), pp. 179–99 (at p. 179).

39 'Christina' is a pseudonym for Maillart's travel companion Annemarie Schwarzenbach.

40 Stamm, "Zwischen Anpassung und Widerstand", p. 185, p. 192.

41 Owyong, "Clothing Semiotics", p. 196.

42 Owyong, "Clothing Semiotics", p. 206.

43 Melissa Kerr Chiovenda, *Agency through Ambiguity: Women NGO Workers in Jalalabad, Afghanistan* (University of Connecticut Graduate School, MA Thesis, 2012), p. 31.

44 Chiovenda, *Agency through Ambiguity*, p. 31.

45 Chiovenda, *Agency through Ambiguity*, p. 31.

46 Annemarie Schwarzenbach, "Der Tschador", in Regina Dieterle and Roger Perret, eds., *Auf der Schattenseite*, pp. 233–35.

47 Nowadays, the word *chador* is still used, but it refers to a full-body cloak which still leaves the face visible. The differences between several types of veiling are explained and visualized online in a BBC Newsround article titled "What's the

Difference between a Hijab, Niqab and Burka?" *BBC Newsround* (7 August 2018), https://www.bbc.co.uk/newsround/24118241 [4 October 2022].

48 Calefato, "Fashion as Cultural Translation", p. 348.

49 Calefato, "Fashion as Cultural Translation", p. 350.

50 Kate Manne, *Down Girl: The Logic of Misogyny* (Oxford: Oxford University Press, 2017), p. 1.

51 Interestingly, although Schwarzenbach's and Maillart's texts on Afghanistan originate in a shared journey, the depictions of the encounter with a European woman wearing a chador in their texts do not match at all, indicating that both accounts do not seem to be based on the same factual encounter.

52 It should be mentioned that the narrators in Maillart's and Schwarzenbach's texts do not blindly support modernity. While taking a positive stance on values such as women's emancipation, they are critical of industrialization and its effects. For a discussion of the paradoxical combination of both Eurocentrism and Eurocriticism in their texts, see Decock and Schaffers, "Die Darstellung der 'anderen Frau' in Reisetexten", pp. 789–91.

53 Anna-Mari Almila, "Introduction. The Veil across the Globe in Politics, Everyday Life and Fashion", in Anna-Mari Almila and David Inglis, eds., *The Routledge International Handbook to Veils and Veiling* (London: Taylor & Francis, 2017), pp. 1–28 (at pp. 6–13).

54 Almila, "Introduction", pp. 1–4.

55 Decock, *Papierfähnchen auf einer imaginären Weltkarte*, pp. 137–39 and pp. 240–520.

IV

Mobility, Perception, Experience

11 Surfing Wanderlust
Surf-Tripping Bodies as Cultural Bearers

Anne Barjolin-Smith

11.1 Introduction

Surf trips are a fundamental element of the surfing experience. As quests for the world's best waves, they represent the quintessence of traveling bodies in that they are the ideal expression of a journey in which the physical body is at once a tool for exploration, performance, communication, and representation. Thus, more than a touristic journey, they are a wanderlust often defined as an affliction.[1] Surf trips have been discussed in terms of traveling bugs stemming from a physical need to feel bodily sensations of pleasure;[2] however, very little has been said about the cultural significance of surfers' bodies during surf trips. Since the nineteenth century, modern surfing culture has rested on myths of progress, frontier, and white male achievements as the United States has built it. This mythology has been fulfilled by a pantheon of Californian surfing legends (erasing the history of Hawai'ian legends), by surf music as the expression of an authentic surf lifestyle developed in California in the 1960s, and by the figure of the American surfer portrayed in mainstream media as a new explorer.[3] Surfing culture is territorial as it invades both bodies and spaces.[4] Based on their corporeal differences, the objective bodies of surfers are determined by their society and culture, which they also express and embody. This way, surfing culture occupies a sensible space, enabling surfers to mark their surroundings through their bodily practices. Thus, the surfer's traveling body is a political figure whose representation is underlined by cultural, ideological, historical, and geographical issues.

This ethnographic study, based on participant observation, shows how Floridian surfers' bodies carry a form of American mythology from their own land to a 'foreign' land, the Maldives. Their Western surfing culture, made manifest in their bodies, is contrasted to a version of Islamic culture through an analysis of surf-tripping bodies as cultural bearers. These surfers' practices mirror their community's complex subjectivities and singularities, thus giving their United States surfing culture a face and voice. Their traveling bodies function as agents of imperialism, diplomacy, and ambassadorship. They also act on collective representations of exogenous

DOI: 10.4324/9781003331803-15

groups by allowing surfers to share their individual experiences in a 'foreign' context. Through their surf trip, the surfers observed yearn for the global as they feel that the local is not enough, yet they differentiate themselves from their hosts in ways explored in this chapter. As a form of play, the surf trip is a way of liberating their physical bodies from their capitalist society's work constraints. However, their journey also subordinates them to their need for bodily pleasure and the cultural charge assigned to their bodies in a 'strange' place.

Because surf trips allow bodies to become instruments of cultural representation that generate ideological meaning and worth, the notion of surf-tripping bodies as physical, mobile, political, cultural, national, glocal, and symbolic representations compels the complex investigation proposed in this chapter. Before embarking on this journey as an observer and participant, I pondered whether there was a specific American body language among surfers. If so, where, when, and how would this language manifest itself? This inquiry, set in the context of a collaboration between the Maldivian and Western cultures, generated several other questions: what would the contrasts between these two cultures in the context of the surf trip reveal about American surf-tripping bodies? How do surfers' traveling bodies function as agents of representation of some fundamental aspects of the US ethos, including imperialism, capitalism, individualism, freedom, and admiration for self-made pioneers? In terms of representation, what was the role of the physical body in the study's multicultural setting? In addition, the underlying meanings of the surf trip had to be interrogated since surf tourism was inherently built as a form of colonizing activity from its very beginning when North Americans discovered surfing in Hawai'i,[5] appropriated the practice, and turned it into the global industry it has become.[6] As a form of colonialism, surf tourism has promoted the importation of Western cultural, social, and economic values, codes, and models, influenced by the United States, in the Global South.[7]

This study is an attempt to answer these questions by combining the notions of the physical body as a multidimensional bearer of culture and the surf trip as an American form of imperialism. Accordingly, I examine how American surf-tripping bodies display American sociocultural practices, history, and ideologies by reproducing Americanness in 'foreign' settings. After defining the functions and meanings of the surfer's body in the fieldwork context, I clarify the issues underlying the terms of an American surf trip to the Maldives. Further, I analyze how American surfers' wanderlust showcases Americanness (the quality of being American) and enables Americanity (American domination founded on the idea of progress). Then, I show how surf-tripping bodies have been agents of imperialism throughout surfing history since it became a North American practice, and I outline the consequences of the surf trip on the host nation.

11.2   The Fieldwork

This transdisciplinary research is rooted in American Studies,[8] and thus, it relies on tools developed by various disciplines to explore aspects of the history and culture of the United States and to look at how they affect social and political realities. I am particularly interested in Anibal Quijano and Immanuel Wallerstein's notion of Americanity, which suggests a form of domination of the Americas (a new world) subordinating other nations in the modern world system.[9] Here, the notion applies to the United States only. Thus, to discuss Americanity, it is necessary to understand what Americanness refers to, which is the quality of being American or being a citizen of the United States. Sports sociology and cultural anthropology theories support the empirical analysis of the body in American surfing culture as represented by Floridian surfers on their trip to the Maldives. As a participant and an observer, I assessed how they expressed Americanness and underpinned Americanity – two intertwined concepts that encapsulate the notions of citizenship, masculinity, whiteness, progress, and coloniality – a common form of domination illustrated by the hegemony of Western cultural canons,[10] here, those of surfing culture.

The international group of travelers was composed of eight Americans (three advanced Floridian male surfers, three male beginners from the west coast of the United States, one Floridian woman who did not surf), and three French people from Reunion Island (one intermediate male surfer, one advanced male bodyboarder, one advanced female bodyboarder), and one man from mainland France who did not surf. The study focused on the three advanced American male surfers who learned surfing and developed their practice in Florida over several decades. The ship's crew was composed of a captain, four crew members (including a chef), two independent surf and dive guides, and an independent *dhoni*[11] captain and his mate. All were from the Maldives, and all were Muslims. While most participants knew about my research project, I decided to minimize my role as an observer to prevent surfers and crew members from modifying their behaviors. Thus, no formal interviews took place, so all the data derives from informal conversations and observations consigned to a journal.

11.3   The Surfer's Body

Conceptions of the body differ immensely depending on how it is looked at. Thus, before exploring surf-tripping bodies as bearers of culture, it is necessary to outline the complexity of the term within a selective range of definitions of the body relevant to this chapter.[12] Beside revealing individuals' age, health, and strength, enabling various types of performances (including surfing), the physical body is multidimensional and is endowed with complex meanings. In the Maussian tradition, it has been considered a social instrument that needs to be learned. Accordingly, as Anthony Synnott and David Howes point out, it is "socially constructed [and means] different things to different people, but very different things to different peoples".[13] Building on this idea, the body has been conceptualized as a symbolic construction culturally and socially defined – a bearer of meaning.[14] In the Foucauldian tradition, the body, which can also be collective and individual, may be invested with political power imposed by cultures and societies, thus embodying them. These dimensions of the body are expressed by its potential vocal and kinetic communicative functions. The body can be at once agent and symbol, objective and subjective, and therefore, it can be envisaged in its corporeality and as embodiment. In all its forms, the body has been represented in various forms of narratives, including myths, and through various media types, including films and social media.

Consequently, in this study, surfers' tripping bodies are at once physical apparatuses trained to behave and communicate according to the US sociocultural codes, political resources used in surfing to display masculinity[15] and whiteness, and symbolic vehicles for the history and the imaginary of the American nation. By occupying a 'foreign' space, surf-tripping bodies become instruments of cultural representation that function as agents of imperialism despite themselves. According to Tara Ruttenberg and Peter Brosius, "surf tourism and its associated development are inextricably conflated with neoliberalism, itself a construct only decades in the making, and its hegemony in defining values, beliefs, practices, and forms of governance".[16] Thus, on their journey, the American surfers under observation bear what I call an intrinsic cultural load that they may not necessarily be aware of. Indeed, there are two ways for travelers to embody their culture: with volition, as they consciously display signs of national and cultural belonging (for instance, through clothing), or without volition, as individuals become archetypes of their nation, culture, and society by proxy, as soon as their background is disclosed. This form of cultural embodiment constitutes their intrinsic cultural load and is the focus of this chapter.

11.4   Surfing as an American Lifestyle

The place of the body in surfing is complex and multidimensional. Exploring and understanding these complexities requires outlining some of the key elements that constitute surfing. Today, an extensive academic literature on surfing has labeled the practice with a variety of terms, including lifestyle, extreme, alternative, or action sport.[17] In this chapter, surfing is understood as a lifestyle sport, a term used by participants themselves to designate a sport that allows them to express a sense of sociocultural identity and to leave the framework of classic sports activities like soccer.[18]

Lifestyle sports are often qualified as extreme,[19] but this epithet refers to an exceptional approach and sensational representation of the practice as it is promoted by the X Games, for instance.[20] Put simply, surfing as a lifestyle is the expression of a global culture constructed and singularized in the local, enabling communities of practice to form around common practices, ethos, aesthetics, and ideologies, which, respectively, could be exemplified by regional approaches to surfing, like the Floridian spirit of fun and positivity, music consumption, and a system of values that differs from so-called achievement sports.

In its post-Hawai'ian tradition – in 1960s California – surfing became a youth movement that enabled young Americans to express their anti-establishment values. Progressively, surfers developed distinct approaches to this lifestyle, ranging from soul surfing to professional surfing as the two opposites of one spectrum. Thus, surfing took on various symbolic meanings over time and space and went from being a traditional element of Hawai'ian culture to being characterized as a subculture or counterculture in its North American iteration. While post-subcultural studies have disputed these two labels, various reassessments of the term 'subculture' have emerged, which are relevant to discussing the specificities of surfing. I suggest that surfing subcultures are regional units that interact with each other to compose the global culture of surfing.[21] Accordingly, while subcultures illustrate local approaches to surfing and retain some distinctive and anti-mainstream characteristics, the global culture of surfing necessarily includes mainstream elements; however, these two conceptions of surfing cannot be envisaged separately. In *The American Surfer*, Kristin Lawler defines surfing (sub)cultural values as "short-term hedonism; spontaneity; ego-expressivity; novelty and excitement; activities performed as ends in themselves; and disdain for work".[22] In order to reflect the evolution of surfing, its diversity, its complexity, and its representations, it is necessary to set Lawler's definition against surfing's passage into the mainstream. For instance, Kyle Kusz highlights that in the 1990s, mainstream magazines such as *Time* "[d]epicted extreme sports as activities which were resurrecting traditional American ideals such as pushing boundaries, taking risks, and being innovative".[23] He suggests that the decade represented surfing in magazines, newspapers, films, etc., through the lens of "the traditional American values (like individualism, self-reliance, meritocracy)".[24] In other words, the United States appropriated the Hawai'ian culture of surfing, then matured modern surfing, reflecting American values of capitalism and individualism of the late twentieth century.[25]

The appropriation of Hawai'ian surfing culture by the United States encapsulates the nation's ethos. First, Americans settled in Hawai'i, then they opportunistically (re-)constructed surfing as a North American practice.[26] Indeed, while surfing was declining in Hawai'i in the nineteenth century, Americans claimed to have revived it while blaming Hawai'ians for *shamefully* letting the practice die. For example, after traveling to Hawai'i in the early twentieth century, Jack London's wife, Charmian London, reiterated Alexander Hume Ford's view that "[t]hey [the Hawai'ians] ought to be ashamed for letting it [surf-boarding] languish".[27] As recent research has shown, throughout the nineteenth and twentieth centuries, American scholars and media relayed an inspirational story of a culture left to perish by its own people, revived by Americans, and turned into an American myth of discovery, transcendence, and conquest – of waves, territories, and inevitably of people.[28] In the United States, surfing went from being a youth countercultural movement to being a representation of reinvention, progress, fun, and white self-made pioneers – a narrative cherished by Americans and widely used by the media and the corporate world as a marketing tool. According to Kusz, "[i]t is within this politico-cultural context which this mainstreaming of extreme sports – which involves its Americanisation, masculinisation, and implicit racialisation – took place".[29] That being the case, surfing became a representation of the American spirit. In turn, surfers' traveling bodies have become ambassadors for their nation and culture by (implicitly) promoting subcultural and mainstream symbols of surfing as a white, masculine, American practice.

11.5   The Political Charge of Surf Trips

The surfer's representation within the context of a surf trip entails more than the physical journey. Surf trips are mythological constructions[30] built over time by the media and the surf industry, starting in the early days of North American surfing, in 1960s California. In its modern form, the surf trip is an American creation that is tributary to its creators' image. As noted by Ruttenberg and Brosius, the surf trip is "a travel-to-surf narrative of tropical surf nirvana to satisfy an insatiable endless summer dream".[31] The surfing world has developed a need among surfers to explore unknown territories with promises of unattained satisfaction and freedom. Surfers want to feel the stoke, a feeling of joy, peace, and connectedness with the ocean.[32] Americans also perceive the surf trip as a way to escape the constraints of everyday life. Surfers observed for this study had very busy lives[33] and had planned for this exceptional experience for months. The paradox was that they liberated themselves from social responsibilities during the surf trip, but their appetite for waves and their need to take advantage of every single one of them soon took over. They approached the surf trip as Americans are taught to engage in sports: uncompromisingly for fun and performance.[34] On this trip, some participants would surf to the point of physical exhaustion and suffering. Some of them did not wish to visit the islands because their corporeal need for sensations – the pleasure induced by the excitement of catching a wave – was stronger than the desire to learn about their host environment. For these surfers, there was no need to think about the context they were in (how to behave and speak in terms that would not offend the locals); the focus was on action based on a reflection regarding the best ways to achieve bodily pleasure – the stoke. Thus, despite the notion of freedom attached to surf trips, as surfers liberate themselves from their social duties, they become subordinate to their own body's urges. However, as Michel Foucault contends, "being free means not being a slave to oneself and one's appetites".[35] He argues that the lack of "ethos of freedom", the practice of freedom, could transpose into a lack of care for others.[36] Thus, the absence of interest, even the sense of superiority expressed by some surfers toward their hosts' culture, suggests that they lacked the ethos of freedom.[37] This behavior is reminiscent of a new form of imperialism and coloniality discussed extensively in surf studies literature in the past 15 years.[38] Indeed, surf trips involve more than just the surfers' stoke and, as instruments of imperialism, are also political.[39] From their early representations in American movies to their implementation, surf trips have raised issues of otherness, legitimacy, and imperialism. Bruce Brown's archetypal surf movie epitomizes the surf trip representation. In *The Endless Summer* (1966),[40] Brown depicts white surfers abundantly catching the world's best waves in a narrative articulating a rhetoric of quest with the myths of the frontier and the naive natives unaware of their surroundings' full potential.[41] The imperialistic tone of the movie relegates locals to primitive and blissful ignoramuses. While these white American surfers travel through politically unstable countries, such as South Africa in the 1960s, the narrator, Brown, never alludes to this. According to Ralf C. Buckley, Daniela Guitart, and Aishath Shakeela, "[m]any of these new surf tourism destinations are in developing nations with no local history of surfing".[42] Therefore, they are developing according to the example set by these new explorers.[43] It is worth noting that it did not take long for these nations to develop a solid surf-oriented tourism industry. These places compete to offer the best conditions for surfers worldwide by westernizing facilities and adopting the American dollar, the English language, and Western surf practices. As a result, surfing and surf trips have become a part of a very dynamic tourism industry. As noted by Buckley, Guitart, and Shakeela, "[d]ifferent sites, known as surf breaks, differ greatly in their attractiveness to surfers. Surf breaks rely on ocean swell exposure and wave climate, and ocean floor topography (bathymetry), to generate a consistent supply of high-quality surfable waves".[44] The surf quality is the reason American surfers are willing to invest a substantial amount of money to go to the Maldives. This is especially true for Floridian surfers, whose beaches' bathymetry does not consistently generate good (surfable) waves.

11.6   Stakes of a Surf Trip to the Maldives

Surf trips usually consist of migrations from the Global North to the Global South, so destinations are geographically and culturally different from the travelers' home country, as was the case for our participants. The Republic of Maldives is an independent nation located in the Indian Ocean. The capital city is Male. The archipelago, comprised of 26 atolls, gained independence from the United Kingdom in 1965, a relationship explaining the Maldivians' mastery of the English language. Before becoming a world-class destination for surfers around the globe, the Maldives were "a tropical island beach destination and a specialist dive tourism destination".[45] Until 2008, the government promoted tourism development based on the principle of "maximum cultural separation between Western tourists and Muslim citizens" (187). While later governments tried to limit this separation, it is still implemented today through the isolation of resort islands, preventing a confrontation between the alcohol-free Islamic nation and Western tourism. For instance, before boarding the ship for the week-long surf trip, some surfers stayed in one of the island-resorts, where alcohol was allowed. As Linda Richter points out: "Although the government generally adheres to Islamic injunctions against alcohol, pork, and graven images, such restrictions do not apply on the resort islands."[46] Every resort is a self-contained island, so tourists do not have to leave them to access food, alcohol, or outdoor activities. Accordingly, surfers stayed somewhat isolated from the authentic Maldivian experience. They went from the resort island to the boat, except for punctual visits to a few islands split between touristic and local zones abiding by Islam law (thus, prohibiting alcohol and compelling women to cover their shoulders in the public space).

Under these somewhat strict conditions, it seems unlikely that the Maldives would become such a popular destination for hedonistic surfers around the globe. However, the Maldives offer an exceptional setting for intermediate to advanced surfers, comprising ten surf breaks in the northern atolls and seven in the South. The surf breaks are all reef breaks that "typically form consistent high-quality waves, but they require high skill to avoid wipe-outs onto shallow coral rock".[47] In contrast, Florida – home of the three advanced American surfers – is all beach breaks with sandy ocean floors. In the Maldives, the reef break is more forgiving than most other popular reef breaks because they "are covered by shelving benches of broken coral fragments which soften the surf break. [...] This is one of its attractions in the international surf tourism market: high-quality reef-break waves, but lower risk of injury for older [or less experienced] surfers".[48] Besides, the Maldives offer two types of surf tourism: an island-based one and a boat-based one. According to Richter, "[t]here is a growing boat-based sector, including luxury cruises and specialized live-aboard charter boats, departing from a dock at the international airport".[49] This option was one of the determining factors in the participants' decision to go to the Maldives because it gave them access to a variety of surf spots and an easier approach to the waves.[50] They had the possibility to move from one surf break to the next on demand, depending on surf conditions.

On the boat, the crew adjusted to their guests' schedule, which meant that they did not pray five times a day, as the call of the muezzin resonating in the atolls would compel them. The two guides drank beer with the surfers and were not bothered by the fact that the three women on board walked around the ship wearing bikinis only. During an informal conversation, they admitted that their extended contact with Westerners had changed their cultural practices. That said, sociocultural contrasts between the Maldivian crew and the American surfers remained observable in the ways their physical bodies displayed everyday know-how and ways of being specific to their sociocultural groups. This phenomenon may be explained by the idea that the social world (in the Maussian sense of determined societies) alters the body so that distinctions among peoples become more obvious through their usage of the body. Thus, according to Synnott and Howes

> [T]hese essentially social processes are also essentially physical, and are marked variously by eating and drinking, or perhaps special clothes and decorations, changes in hair, ritual cleansing, body-painting or cicatrization, body-mutilations of various sorts: tooth removal, circumcision, sub-in-cision, and so on, with the rites and symbolism varying from culture to culture.[51]

While several societies may share many processes, it is the combination of a set of processes that singularizes a social group. In addition, the body is also learned differently according to cultures, and surfing is an example. Indeed, besides their typical garments and bleached hair, global surfers can be recognized by how they stand as their upper body may be more developed and their back more arched than non-surfers. American surfers can be recognized by how they eat, speak, and interact. Thus, the combination of social and cultural processes embodied by surfers betrays their sociocultural heritage. These behaviors are not intrinsic and have to be learned; hence, they are subjective and carry the meaning of individuals' time, culture, and society. Aesthetic criteria vary from one group of people to the next, and they also vary in time, thus telling individual and collective stories. The processes in question have a lot to reveal about these traveling groups of people. They are the first step to understanding the meaning that their bodies bear as they travel, encounter other cultures, and occupy their space.

11.7   Surfing Wanderlust: Interpreting the American Surfer's Body

This investigation of the traveling body's meaning as a bearer of culture raises questions of representation, and in our case, of what it means to be an American surfer. In previous research on the ethno-aesthetics of surf in Florida,[52] I interviewed surfers who claimed to be American surfers based on their technique and their attitude toward surfing and other surfers. There is a sense of being American as a surfer, and to understand the singularities of these surfers' bodies as American bodies, we must question the processes involved in learning first to be an American, then to be an American surfer. How is the sense of being American and belonging to this nation conveyed by the body? How does the duality between *Us* (Americans) and *Them* (non-Americans) convey a feeling of *Us, united* and *Us, singular individuals*? America is composed of a multitude of people, communities, beliefs, etc., so how can we reconcile this diversity into a standard American culture and society?[53] Arguably, the unity of the American people may be found in their shared history, mythology, and future. In her work, *Male and Female*, Margaret Mead discussed the seemingly impossible task of comprehending the American people as a whole because of the complexity of its composite culture.[54] Nevertheless, despite the nation's sociocultural diversity, she identifies some characteristics that differentiate the American people from the rest of the world. She argues that people in America may behave and live differently; however, Americans behave and act American: they speak the same language, they are future-oriented, and they share a common imaginary, or dream, of what it means to be an American. Mead contends that "the dream takes precedence" over individuals' reality.[55] In other words, being American means building and protecting a collective imaginary of the US ethos based on the nation's historical and mythological constructions (including a constant reinvention of Manifest Destiny, capitalism, individualism, etc.). The commonality of the dream Mead refers to has been tied together and been reinforced by an essential and multidimensional corporeal function: language. In its various acceptations as speech, identity representation, kinesics, cultural marker, and form of imperialism, a language learned by the body is an expression of sociocultural and political heritage. Americans have developed a distinctive form of English endowed with identifiable linguistic features and a characteristic loudness. Linguistic variations within American English translate regional identities. Kinetically speaking, American's body language tends to promote personal space, and some specific hand gestures, such as air quotes or the Hawai'ian *shaka*, may convey their origins. As a cultural marker, American English has become the *lingua franca* of the global surfing world, a shift from Hawai'ian to English, otherwise called linguistic imperialism, comparable to the Maldivian situation with British English. These shifts echo Arjun Appadurai's description of the relationship between Bombay and England

and the United States: the Maldives seems to have moved from a British postcolonial subjectivity to another American-centric one.[56]

This form of imperialism and the commonality of the American imaginary described thus far are central to Americanness and Americanity. However, these notions' implications have evolved since the advent of surfing as a global practice in the 1960s. The surfing world has taken on new forms of imperialism, exploitation, hierarchization, and domination dictated by capitalism, narrated globally in English, and relayed by American media. White males have been photographed and filmed in the ocean, conquering waves and constantly crossing physical and mental limits.[57] For Kyle Kuzs, "[t]he frontier which challenges, yet is ultimately conquered by, white American men is a foundational national trope which reproduces not only dominant ways of how America imagines itself, but also how American manhood is imagined".[58] The notion of frontier is fundamental in the construction of the American mythology. Whether it means exploring unknown territories on a surf trip or pushing one's limits by taking risks, for Kusz, the place of frontiers in extreme sports, such as surfing, makes them a current representation of the American mythology: "Both of these white male figures – the extreme sportsman and American frontiersman – are valorised as ideal American citizens because they are invested in a set of distinctly American values: rugged individualism, self-reliance, personal responsibility, and individual/social progress."[59] The notion of progress is fundamental as it is anchored in America's Manifest Destiny. As Mead points out, Americans are "oriented towards the future";[60] however, while they look toward the future, they also ground their identity in the construction of an epic past. Americans are on a constant quest for new frontiers, literally and figuratively, but this should not be construed as their neglect of a mythological past. The myth of the American identity is that of the explorer pushing back the boundaries – it is a wanderlust. Americans go West, to the moon, and on surf trips. The myth of surf trips dovetails the idea of a mythical frontier passed. In this wanderlust, the American body – collective and individual – is at work: a *We the people*, and an *I individual, able and free*.

This representation of Americanness is intrinsically and historically white, ultimately impacting the relationship between the *Us, Americans*, and the *Them, Non-Americans*. In his essay on the content of whiteness, Nelson M. Rodriguez examines the discourses of whiteness, identifying three rhetorical strategies.[61] He argues that, firstly, there is a scientific classification of whiteness, which eludes social questions and power relations. Secondly, nationality is conflated with race so that *White equals American*, an essentialist view that controls identity and marginalizes anyone who does not match *White equals American*. Additionally, this particular ideology removes from memory and history the contribution of non-whites (including non-masculine, non-heterosexual figures) to Americanness to "maintain asymmetrical power relations".[62] Third, whiteness is typically understood in relation to European ancestry, removing any duty to critically self-reflect on whiteness. Accordingly, these rhetorical strategies suggest that surfing as Americanness places non-whites on the margins of the practice, including Hawai'ians, as it has become a cultural space for whiteness. However, as Kusz expounds, the paradox is that most white participants and mainstream American media seldom envisage surfing as racially exclusive.[63] This is how Americanness supports a singular practice of Americanity in which a dominant people has taken over and renewed a culture that was bound to disappear – or so it was claimed in the case of Hawai'ian surfing. The American press has relayed this representation of surfing beyond America's frontiers and has helped solidify a notion of Americanness embodied by surfers.

11.8   The Embodiment of Americanity in Surfing

Americanity, as a concept, rests on the idea of newness as a means of domination. According to Quijano and Wallerstein, "[t]he newnesses were four-fold, each linked to the other: coloniality, ethnicity, racism, and the concept of newness itself".[64] Coloniality represents the dominance of a state over other states within a hierarchical system, and even with the end of colonies, coloniality remains "in the form of a sociocultural hierarchy".[65] Americanity had a role in shaping the global surfing world, with the United States as its instigator. Within the ethnicity aspect of Americanity lies the idea that each ethnic group is assigned a task and a place in society. Coincidentally, in the context of our surf trip, tasks repartition on the boat took on a stereotypical turn based on skin color repartition: dark-skinned were hosting while white-skinned were leisuring; a separation of roles in surfing media representation that has been the unfortunate norm since the release of Brown's movie. As a result, while most American surfers denounce racism, it is nevertheless latent in American surfing through unquestioned and tacit practices of exclusivity. The moral denial of an exclusive reality is connected to the concept of newness defining Americanity. It is based on modernization processes and involves a lack of acceptance or recognition of the dominant culture's fault within the process. For instance, while the surfing world claims to be inclusive and ecologically minded, it is, in fact, exclusive, and its development rests on its ability to pollute,[66] most likely at the expense of subordinated nations. In the Maldives, while surfers consumed bottled water and witnessed the decay of the Maldivian atolls,[67] it was upsetting for them to accept that they were a part of the problem. It is also morally uncomfortable to acknowledge that surf trips follow one pattern: from the hegemonic country to a subordinated one. Surfers travel South for bodily pleasures and go to places from which people travel

North for survival. The character and implementation of Americanity rest on the negotiation of all these paradoxical elements and compel us to look at how Americanity constitutes an intrinsic cultural load for surfers, a load they may embody.

Interpreting the American surfer's traveling body against the backdrop of Americanness and Americanity underlines issues regarding the representation of surfers' traveling bodies within a 'foreign' place they have involuntarily taken over in several ways (by imposing their surfing culture and their language and by indirectly transforming the local industry to cater to their needs). The Maldivian culture and American surf culture are complete opposites in terms of body representations. The American surfers bore the legacy of Tom Blake, the prototype of the American surfer, whose "sun-bleached, rumpled personal style became the prototypical beachcomber look, still in effect today".[68] Their bodies had been socialized through the way surfing had modified them and through their agency (their involvement and control of their surfing). As Synnott and Howes note, "[t]he senses are trained to operate in socially approved grooves".[69] Individuals learn to behave according to established codes validated by their sociocultural group. However, these practices may be challenged when socialized bodies are displaced and put in a different society. In the context of the surf trip, bodies shift from being symbolically charged to being politically charged. Individual and collective bodies engage in a power struggle in which dominance rests on the concept of Americanity's constant move to newness. Synnott and Howes insist that "[t]he body is, after all not only symbolic of the self, and the society, it *is* the self. We are embodied. Furthermore bodies collectively are not only constitutive of the society, they are (or we are) the modes through which society reproduces itself – physically and figuratively".[70] Participants represented their society not only because they looked American but because they embodied Americanness, reproduced their society physically and figuratively, and were associated with a certain imaginary of the United States: a surfing America which Laderman describes as "a cultural force born of empire, [...] reliant on Western power, and invested in neoliberal capitalism".[71]

11.9 The Surf-Tripping Body as Agent of Representation and Imperialism

The ideological, cultural, and geographical issues that underlie the representation of the American surfer's body are anchored in Americanness and Americanity. These two notions rest on the ideas of frontier, whiteness, and masculinity and unveil the myth of the white American male driven by wanderlust. As they travel, these surfers represent their nation despite themselves and reproduce patterns of imperialism intrinsically attached to their Americanness, present, past, and future. America's antagonizing circumstances on issues of race, social, and gender equality place white masculinity as the dominant yet threatened figure of the United States. As representations of their nation, traveling surfers in the twenty-first century have been embodiments of the rejection of Islam, 'foreigners', and female emancipation idealized by the conservative fringe of the population and relayed by former President Donald J. Trump on the world stage. This representation of the archetypal white American male has gained momentum as it has built upon previous instances in various communities of practice (e.g., 1990s whiteness and masculinity in extreme sports, whiteness in American country music). The surf trip is, by essence, constructed and coded as white American masculinity because of its history and media exposure across America and the globe. The fact that all American surfers on the boat were white and blond with blue eyes may sound anecdotal, but it resonates with the current dominating embodiment of the American stereotype. According to Kusz, extreme sports afford white males a sense of control over their own masculinity within their society. Their exploits must be recorded, displayed, and promoted. Iconography is essential in surfing as it tells its American history, its ethos, and its evolution. Recordings allow surfers to create tangible memories, inundate the rest of the world with proof of their actions, claim a premium place in history, and preserve it. As Lawler observes: "That's why the images have proven so inspirational, and have always had a privileged place in the American imagination."[72] The surfer's image and the representations of the lifestyle all work together to build the American figure of the surfer. According to Serge Gruzinski, images have accompanied the processes of colonization of indigenous imaginaries because they have had the power to convey meaning across linguistic barriers: "[images are] used as a marker [and] a tool for acculturation and domination".[73] By occupying the Maldivian space, Western surfers' culturally loaded bodies leave a distinct mark on the youth of the atolls who come into contact with them. They co-opt them into their Western expression of identity and become instruments of rebellion for the locals. For instance, the surf and dive guides explained that their generation was moving away from a rigorous approach to Islam. In a country where the call of the muezzin resonates throughout the atolls five times a day, these 25-year-old men drink, smoke marijuana, do not pray, and are not concerned about exposed women's bodies. Their behavior contrasts with the local customs. In fact, during a visit to an island, these guides were summoned to tell women to cover their shoulders. Thus, the cultural gap between Western and Maldivian cultures seems to be narrowing. According to the surf guides, this is due to increasing Western tourism and the fact that the Maldivian youth is now connected to the internet, specifically to Instagram, the epitome of modern mythological image construction. Surfers' bodies become images synthesizing what it means to be American. Their bodies are transformed into raw material for social media content that surfers themselves create and disseminate globally. While the role of photographs, artwork, cartography, and writing has been analyzed in sports literature to show how the United States has imposed its cultural power,[74] this contribution has illustrated that the surfer's multidimensional body has a similar power.

In sum, the context of the surf trip put in perspective a nation's portrayal by confronting perceptions of *Us* and *Them* as actualized preconceived mythologies. Participants embodied the myth of the prototypical American surfer, singularly portraying the American nation. Overall, the cultural representation endorsed by surfers is not necessarily voluntary but intrinsic. Ultimately, it makes them the bearer of an American culture whose characteristics they do not always agree with or implement but that they embody in the eyes of the other, which might simply be a matter of cognitive economy (a mental simplification of categories illustrated by stereotypes). In a comparable setting involving surfers from other countries (for instance, French surfers in Hawai'i), we can expect similar results regarding how cultural differences may contrast, emphasizing how bodies truly and clearly are bearers of cultures. The exploration of the meaning of the surfer's body adds a new dimension to the study of surfing and the figure of the American surfer, but also to the study of the body's function in global cultures and societies.

Notes

1 Scott Laderman, *Empire in Waves: A Political History of Surfing* (Berkeley: University of California Press, 2014), p. 43.
2 Mark Stranger, *Surfing Life: Surface, Substructure and the Commodification of the Sublime* (Burlington: Ashgate, 2011), p. 122; and Meg Samuelson, "Searching for Stoke in Indian Ocean Surf Zones: Surfaris, Offshoring and the Shore-Break", *Journal of the Indian Ocean Region*, 13, no. 3 (2017): 311–25.
3 Anne Barjolin-Smith, *Ethno-Aesthetics of Surf in Florida: Surf, Musicking and Identity Marking* (Hampshire: Palgrave Macmillan, 2020).
4 See, for example, David Lanagan, "Surfing in the Third Millennium: Commodifying the Visual Argot", *The Australian Journal of Anthropology*, 13, no. 3 (2002): 283–91; Richy Bennett, *The Surfer's Mind: The Complete, Practical Guide to Surf Psychology* (Torquay: Griffin Press, 2004); Jon Anderson, "Surfing between the Local and the Global: Identifying Spatial Divisions in Surfing Practice", *Transactions of the Institute of British Geographers*, 39, no. 2 (2014): 237–49; Marilia M. Bandeira, "Territorial Disputes, Identity Conflicts, and Violence in Surfing", *Motriz: Revista de Educação Física*, 20, no. 1 (2014): 16–25; and Lindsay E. Usher, "'Foreign Locals': Transnationalism, Expatriates, and Surfer Identity in Costa Rica", *Journal of Sport and Social Issues*, 41, no. 3 (2017): 212–38.
5 Scott Laderman explains that "modern surfing was born from a conquest" after the United States unlawfully annexed Hawai'i in 1898 (*Empire in Waves*, p. 7).

6 See, for example, Noenoe K. Silva, *Aloha Betrayed: Native Hawaiian Resistance to American Colonialism* (Durham: Duke University Press, 2004); Jim Nendel, "Surfing in Early Twentieth-Century Hawai'i: The Appropriation of a Transcendent Experience to Competitive American Sport", *The International Journal of the History of Sport*, 26, no. 16 (2009): 2432–46; Laderman, *Empire in Waves*; Andrew Warren and Chris Gibson, *Surfing Places, Surfboards Makers: Craft, Creativity, and Cultural Heritage in Hawai'i, California, and Australia* (Honolulu: University of Hawai'i Press, 2014); and Jérémy Lemarié, "Genèse d'un Système Global Surf Regards Comparés des Hawai'i à la Californie: Traditions, Villes, Tourismes, et Subcultures (1778–2016)", PhD Thesis, Université Paris Ouest Nanterre La Défense, 2016.

7 Unlike Tara Ruttenberg and Peter Brosius, I do not think this has created cultural homogenization of the host spaces, even though I concede that these loci respond to "the demands of hypermobile, predominantly white, male, middle-class surfers from the Global North" (Tara Ruttenberg and Peter Brosius, "Decolonizing Sustainable Surf Tourism", in Zavalza Hough-Snee and Alexander Sotelo Eastman, eds., *The Critical Surf Studies Reader* (Durham: Duke University Press, 2017), pp. 109–32 (at p. 110)).

8 In this chapter, "America" refers to the United States and "American" to people from the United States.

9 Anibal Quijano and Immanuel Wallerstein, "Americanity as a Concept, or the Americas in the Modern World-System", *International Journal of Social Sciences*, no. 134 (1992): 549–57 (at p. 549).

10 Luis Chávez and Russell P. Skelchy, "Decolonization for Ethnomusicology and Music Studies in Higher Education", *Action, Criticism, and Theory for Music Education*, 18, no. 3 (2019): 115–43.

11 A *dhoni* is a traditional Maldivian fishing boat.

12 Holly Thorpe has explained how the notion of the body has been disputed in sociology (*Snowboarding Bodies in Theory and Practice* (Hampshire: Palgrave-Macmillan, 2011)). Likewise, Anthony Synnott and David Howes's comprehensive synthesis of the anthropology of the body helps grasp the difficulty of conceptualizing the body ("From Measurement to Meaning: Anthropology of the Body", *Anthropos*, 87 (1992): 147–66).

13 Synnott and Howes, "Anthropology of the Body", pp. 147–48.

14 Mary Douglas, *Natural Symbols: Explorations in Cosmology* [1973], 4th ed. (New York: Routledge, 1996).

15 According to Clifton Evers, surfers "learn to do masculinity" through the role their feelings and bodies play in their sense of belonging as men who surf ("How to Surf", *Journal of Sport and Social Issues*, 30, no. 3 (2006): 229–43 (at p. 235)).

16 Ruttenberg and Brosius, "Decolonizing Sustainable Surf Tourism", p. 111.

17 See, for example, Robert E. Rinehart and Synthia Sydnor, eds., *To the Extreme: Alternative Sports, Inside and Out* (Albany: State University of New York, 2003); and Belinda Wheaton, ed., *Understanding Lifestyle Sports: Consumption, Identity and Difference* (London: Routledge, 2004).

18 See, for example, Wheaton, ed., *Understanding Lifestyle Sports*; Emily Coates, Ben Clayton, and J. Barbara Humberstone, "A Battle for Control: Exchanges of Power in the Subculture of Snowboarding", *Sport in Society*, 13, no. 7–8 (2010): 1082–101; and Robert E. Rinehart and Holly Thorpe, "Alternative Sport and Affect: Non-Representational Theory Examined", in Belinda Wheaton, ed., *The Consumption and Representation of Lifestyle Sports* (Abingdon: Routledge, 2013), pp. 212–35.

19 Douglas Booth, "Surfing: From One (Cultural) Extreme to Another", in Wheaton, ed., *Understanding Lifestyle Sports*, pp. 94–110; George Eric Brymer, "Extreme Dude! A Phenomenological Perspective on the Extreme Sport Experience", Ph.D. Thesis, University of Wollongong, 2005; and Holly Thorpe, *Snowboarding Bodies in Theory and Practice* (Hampshire: Palgrave Macmillan, 2011).

20 See Holly Thorpe and Belinda Wheaton, "'Generation X Games', Action Sports and the Olympic Movement: Understanding the Cultural Politics of Incorporation", *Sociology*, 45, no. 5 (2011): 830–47.

21 Barjolin-Smith, *Ethno-Aesthetics of Surf in Florida*.

22 Kristin Lawler, *The American Surfer: Radical Culture and Capitalism* (New York, London: Routledge, 2010), p. 3.

23 Kyle Kusz, "Extreme America: The Cultural Politics of Extreme Sports in 1990's America", in Wheaton, ed., *Understanding Lifestyle Sports*, pp. 197–213 (at p. 197).

24 Kusz, "Extreme America", p. 202.

25 Craig B. Snyder, *A Secret History of the Ollie*, vol. 1: The 1970s (Delray Beach: Black Salt Press, 2011); and Lemarié, "Genèse d'un Système Global Surf".

26 In *Empire in Waves*, Laderman deconstructs accounts claiming that surfing had been "resurrected by the recently arrived haole" (at p. 18). A *haole* is a foreigner in Hawai'ian. It especially refers to a white person.

27 Valery Noble, *Hawaiian Prophet: Alexander Hume Ford: A Biography* (New York: Exposition-Banner Book, 1980), p. 78.

28 Until the late 1990s, English-speaking accounts, including journal articles and tourism ads, accredited the resurgence of Hawai'ian surfing to the promotion of the islands and the sport as a new tourist destination and attraction that started at the beginning of the twentieth century. Several scholarly writings have since shown how the practice never went extinct and had been kept alive by Hawai'ians. The works of Isaiah Helekunihi Walker, Noenoe K. Silva, and Laderman are paramount in showing that the Hawai'ians had been resisting space invasion and cultural destruction by preserving their cultural traditions, including surfing and *mele*. See, for example, Silva, *Aloha Betrayed*; Isaiah Helekunihi Walker, *Waves of Resistance: Surfing and History in Twentieth-Century Hawai'i* (Honolulu: University of Hawai'i Press, 2011); Laderman, *Empire in Waves*; and Lemarié, "Genèse d'un Système Global Surf".

29 Kusz, "Extreme America", p. 202.

30 As a mythological construction, surfing is endowed with a series of mythical narratives connected to its place of origin (Hawai'i) as well as its place of reinvention (California). Surfing's mythology has been constructed by American media (through music, magazines, and films) as intrinsically American. The essential function of these myths is to give a genealogy of prestigious ancestors to the United States who have appropriated the culture.

31 Ruttenberg and Brosius, "Decolonizing Sustainable Surf Tourism", p. 111.

32 Joan Ormrod, "Expressions of Nation and Place in British Surfing Identities", Ph.D. Thesis, Manchester Metropolitan University, 2007; Lawler, *The American Surfer*; and Laderman, *Empire in Waves*.

33 For instance, among the three advanced surfers, one was responsible for the regional branch of a sunscreen company, one was a pharmaceutical representative, and one was a coast guard.

34 Fabien Archambault and Loïc Artiaga, "Plus vite, plus haut, plus riche: La médiatisation de la culture sportive américaine au xxe siècle", *Le Temps des médias*, 9, no. 2 (2007): 137–48.

35 Michel Foucault, "The Ethics of the Concern of the Self as a Practice of Freedom", in Paul Rabinow, ed., *Ethics: Subjectivity and Truth*, trans. Robert Hurley (London: The Penguin Press, 1997), pp. 280–301 (at p. 286).

36 Foucault, "The Ethics of the Concern of the Self", p. 287.

37 It is worth noting that some of the American participants tried to blend with the locals. Indeed, as white American males, they made contact and promoted openness through behaviors and actions that showed interest in their hosts, including learning their names and asking about their relationship with alcohol and tourism as a Muslim nation. They differentiated themselves from the stereotypical American (surfer) characterized by a lack of interest in other cultures and a sense of superiority vis-à-vis non-Western cultures (Hough-Snee and Eastman, eds., *Surf Studies Reader*). They were interested in visiting the islands, and they attempted to create a space of cultural respect and exchange by learning about the culture of their hosts and trying to share the common space on the ship.

38 Colleen McGloin, "Surfing Nation(s) – Surfing Country(s)", PhD Thesis, University of Wollongong, 2005; Laderman, *Empire in Waves*; and Hough-Snee and Eastman, *Surf Studies Reader*.

39 Laderman, *Empire in Waves*.

40 Bruce Brown, *The Endless Summer*. Film. Aviva International, 1966.

41 Informal conversations with surfers have shown that these aspects of the movie are unnoticed or considered irrelevant.

42 Ralf C. Buckley, Daniela Guitart, and Aishath Shakeela, "Contested Surf Tourism Resources in the Maldives", *Annals of Tourism Research*, 64 (2017): 185–99 (at p. 186).

43 Peter Donnelly and Kevin Young explain that neophytes enter a subculture, such as surfing, by imitating members of the subculture and trying to reproduce stereotypes. Neophytes build a new identity by first "distancing from nonmembers", and second by "having the identity confirmed by actual members of the subculture" ("The Construction and Confirmation of Identity in Sport Subcultures", *Sociology of Sport Journal*, 5, no. 3 (1988): 223–40 (at p. 224)). As the rest of the chapter shows, in the Maldives, the contact between Western surfers and Maldivian youth generates the same phenomenon at the individual and the collective levels.

44 Buckley, Guitart, and Shakeela, "Contested Surf Tourism", p. 186.

45 Buckley, Guitart, and Shakeela, "Contested Surf Tourism", p. 187.

46 Linda Richter, *The Politics of Tourism in Asia* (Honolulu: University of Hawai'i Press, 1989), p. 187.

47 Richter, *The Politics of Tourism*, p. 189.

48 Richter, *The Politics of Tourism*, p. 189.

49 Richter, *The Politics of Tourism*, p. 189.

50 The boat rental included a *dhoni* on which the surfboards were loaded and which took the surfers directly to the surf break. Instead of having to paddle through the breaking waves to get to the line-up (the space where surfers wait for a wave to come), surfers were dropped directly next to it by the *dhoni*.

51 Synnott and Howes, "Anthropology of the Body", p. 154.

52 As a concept, ethno-aesthetics encapsulate the way the subjective experiences of a cultural community beget singular forms of expression that articulate sociocultural belonging through dynamics of identity marking and exo-/endogroup representations (Barjolin-Smith, *Ethno-Aesthetics of Surf*).

53 In previous research, I asked the same question about surfing, and I showed that, while there was a global surf culture, it was constructed locally by regional surfing subcultures.

54 Margaret Mead, *Male and Female* [1949] (New York: Harper Collins, 2001).
55 Mead, *Male and Female*, p. 239.
56 Arjun Appadurai, *Modernity at Large: Cultural Dimension of Globalization* (Minneapolis: University of Minnesota Press, 1996), p. 2.
57 During the trip, capturing images of bodies in action became a very important part of the experience. Hundreds of still pictures and motion pictures were taken with various devices ranging from cell phones to GoPros, professional cameras, and drones. Pictures were then displayed at night on a big screen in the boat. The point was to see oneself in action, and very few pictures were, in fact, just about the place. Whether it was within the city or on the water, the focal point remained the participants themselves.
58 Kusz, "Extreme America", p. 203.
59 Kusz, "Extreme America", p. 203.
60 Mead, *Male and Female*, p. 234.
61 Nelson M. Rodriguez, "Emptying the Content of Whiteness: Toward an Understanding of the Relation between Whiteness and Pedagogy", in Joe L. Kinchelo et al., eds., *White Reign: Deploying Whiteness in America* (St. Martin's Press, 1998), pp. 31–62 (at pp. 45–46).
62 Rodriguez, "Emptying the Content of Whiteness", p. 47
63 Kusz, "Extreme America", p. 206.
64 Quijano and Wallerstein, "Americanity as a Concept", p. 550.
65 Quijano and Wallerstein, "Americanity as a Concept", p. 550.
66 Building surfboards is a polluting activity that requires the use of resins and foams (Lauren L. Hill and J. Anthony Abbott, "Surfacing Tension: Toward a Political Ecological Critique of Surfing Representations", *Geography Compass*, 3, no. 1 (2009): 275–96). Engaging in surf trips pollutes since it usually requires traveling by plane. The surf fashion industry pollutes by mass-producing seasonable clothes and attires.
67 Tap water is not fit for consumption in the Maldives. Thousands of plastic bottles were piled up on some islands or burned in the open air.
68 Lawler, *The American Surfer*, p. 71. Tom Blake also famously claimed "Nature=God," not a Muslim or Christian god, but surfing itself.
69 Synnott and Howes, "Anthropology of the Body", p. 160.
70 Synnott and Howes, "Anthropology of the Body", p. 163.
71 Laderman, *Empire in Waves*, p. 7.
72 Lawler, *The American Surfer*, p. 108.
73 Serge Gruzinski, *Images at War: Mexico from Columbus to Blade Runner (1492–2019)* [1990], trans. Heather MacLean (Durham: Duke University Press, 2001), p. 2.
74 See, for example, Laderman, *Empire in Waves*; and Hough-Snee and Eastman, eds., *Surf Studies Reader*.

12 Traveling Bodies in Film

Embodied Encounters and Negotiating Selves

Anne von Petersdorff-Campen

12.1 Introduction

While travel might appear to us first as an activity during which we physically move our bodies through places, it has an impact on us beyond this geospatial dimension. As any embodied experience, it influences our sense of self and personal development overall. Drawing on Maurice Merleau-Ponty, my approach to embodiment is guided by the notion that consciousness is inescapably situated in the body and in the way the body is relating to the world around it.[1] In travel, acknowledgment of our bodies' situatedness has the potential to be heightened. We are sensitized for our "being-in-the-world",[2] and the way we are reciprocally established by the encounters we have and the way we relate our perceptions to our body: "There is an almost quaint correlation between what is in front of our eyes and the thoughts we are able to have in our heads: large thoughts at times requiring large views, new thoughts new places."[3]

As illustrated by Alain de Botton's quote, movement through geographically unknown places can inspire new ideas and mindsets, which might be more difficult to assess when we stay in familiar surroundings. While encountering new landscapes, cultures, and people could be accomplished without leaving our familiar environments, there is something about embodied encounters with the unfamiliar that prompts promising new forms of mental movement. Eric Leed points to the importance of physical movement through spaces in order to develop a set of intellectual skills, like observation and sensing distances, but also to cultivate subjectivity:

> The mental effects of passage – the development of observational skills, the concentration on forms and relations, the sense of distance between an observing self and a world of objects perceived first in the materiality, their externalities and surfaces, the subjectivity of the observer – are inseparable from the physical conditions of movement through space.[4]

This link between the embodied experience and the impact this experience has on our sense of self and our encounters with others marks the

DOI: 10.4324/9781003331803-16

starting point for my investigation. Setting out to explore the dynamics of embodied encounters of women travelers in unfamiliar terrain, I turn to film. My concern with film *and* gender in regard to the role of bodies and embodied experiences in negotiating between self and others is twofold: on the one hand, I draw on film as a way to document and explore the complex mediation between bodies and unfamiliar contexts. On the other hand, I explore filmic strategies and techniques to translate these embodied experiences and encounters to an audience.

I set out on this quest by means of a hybrid endeavor, combining film-making and scholarship. Too often, the professional worlds of academia and filmmaking place practitioners and theorists in separate categories; I would like my work to be understood as an effort to bridge this divide. In doing so, I hope to continue the conversations created by scholars such as Trịnh Thị Minh Hà,[5] Alexandra Hidalgo,[6] Jacqueline Rhodes,[7] bonnie lenore kyburz,[8] and Jody Shipka,[9] who have explored video's richness in their scholarship and demonstrated how moving beyond alphabetic texts can help us make sense of the various intersections between rhetoric, gender, race, culture, and sexuality. Exploring my own travel experience through film, I want to draw attention to the ways in which a shift towards cinematic rhetoric can enrich our scholarship in those fields in general, but *especially* in the context of encounters with other cultures and foreign places. My writing about the filmic staging of embodied encounters of women travelers is then also influenced by co-directing and producing my own documentary, *Wanderlust, cuerpos en tránsito* (2017), together with Maria Pérez Escalá.[10] The film is the result of a journey Maria and I undertook between May and July of 2014, between Egypt and Germany, and documents our travel experiences as women across three continents – Africa, Asia, and Europe.[11]

The experience of traveling as a woman stirred my interest in examining the role of bodies, their perception, their representation, and their interaction with unfamiliar surroundings. A body cannot appear to us outside of a framework, a mode of representation, or a historical and cultural world. Women's bodies are often considered inadequate or out of place, they are perceived as objects of desire or exoticized for the purpose of spectacle, and these representations conversely define and limit where and when female bodies can appear in public spaces and the appropriate time and way these appearances are deemed legitimate. As aesthetic representations circulate in and out of societies and govern to a large degree what is observable, they have political stakes: as ways of seeing and showing, aesthetic representation not only determines what is visible but also what can be said about what is visible. In her essay "Green Screen: The Lack of Female Road Narratives and Why it Matters" (2013), for example, Vanessa Veselka illustrates how the representation of women travelers often produces rigid imaginative frameworks for women and reproduces these frameworks for the societies in and out of which they travel. Prompted by her own experience of hitchhiking and her investigation into the murdering of female hitchhikers in the USA, she summarizes the consequences of limiting cultural narratives for women travelers: "I am not saying that rape and violence aren't the predominant experiences of female hitchhikers; rather, I'm saying that our cultural imagination plays a role in *why* it is the predominant experience."[12]

While there is much to be said about the increased vulnerability of female traveling bodies and the degree to which some mainstream representations may exacerbate these vulnerabilities, the context of travel may also open up rigid frameworks. In unfamiliar terrain, power relations are often rendered more fluid and unstable than in familiar settings. The cultural grid of bodily identity markers, such as gender, race, age, sexuality, economic, and social status, as well as the construction of other bodies in relation to the self, exists in constant negotiation and re-negotiation with a changing environment. On our two-and-a-half-month-long journey across 14 countries, for example, each place surfaced something specific in Maria and me, and also between us, because each place implicated us differently within it. In Egypt, for example, I was made acutely aware of my whiteness, in Israel I became more concerned with my German history and legacy, and the divided island of Cyprus evoked strong associations with the past division of my home country, Germany.[13]

Taking as a point of departure the notion that bodies ground us in time and place and must be scrutinized as central instruments in negotiating constructions of the 'other' and the self, I focus on filmic strategies that highlight the physical and social dynamics of embodied encounters. I engage in this process by means of a three-dimensional way.

The first dimension is a theoretical–historical engagement with literature from feminist film studies. Here, I explore strategies of embodied filmmaking with a focus on the camera-holding body and the voice-generating body as well as the 'footprint' bodies leave on the filmic medium. Accordingly, when I talk about bodies and their role as central instruments in filmmaking, I do not just refer to the bodily image on a screen, but also to the way bodies are involved in the filmmaking process as camera operators and the way they inscribe themselves on an audible level. Subsequently, I explore how audiovisual accounts can affect an audience and help to perform the experience of traveling in a visceral way. In my exploration of embodied filmmaking, I draw on Vivian Sobchack and her reference to Maurice Merleau-Ponty which suggests film experience as bodily intersubjectivity, as an "expression of experience through experience".[14] Referencing this body-oriented phenomenology of film, I explore how representing travel experiences through film can help to negotiate the self in a reciprocally constitutive way with unfamiliar surroundings and in encounters with others.

The second dimension connects the theory with our film, to illustrate the idea that film is itself like a 'lived body' and can help us to make sense of our 'lived body' by exploring and cultivating our own situatedness in the world.[15] I draw on our filmmaking process in *Wanderlust* to offer practical examples of and theoretical insight into how the traveling body inscribes itself on the level of filmic image and sound, exploring the potential of embodied film and filmmaking in the context of travel documentaries of women filmmakers.

The third dimension then involves an analysis of two German autobiographical documentaries that afford examples of women negotiating their embodied selves via film in the context of travel. Through the exploration of selected scenes of Ruth Beckermann's *Eine flüchtige Reise nach dem Orient* (1999)[16] and Helke Misselwitz's *Tango Traum* (1985),[17] I provide examples of how filmic strategies such as haptic visuality or embodied voice-over are used to negotiate and express embodied encounters and the filmmaker's subjectivity.

12.2 Strategies of Embodied Filmmaking

Cinematography, stemming from the Greek words *kinema* (movement) and *graphein* (to write, to record), suggests that the relationship between camera work and the body is concerned with the recording of the movement of bodies and/or the role of the body in recording movement. Expressing and thematizing, rather than concealing, the camera-holding body is associated with cinema *vérité*, which first materialized within earlier filmmaking traditions like avant-garde filmmaking and later resurfaced in genres like the essay film and experimental filmmaking.[18] Drawing on these traditions, film scholars have (re)focused their attention on the physical involvement of bodies in the act of creating cinematographic images. Michael Albright, for example, describes "cinematographic embodiment" as the "instances when the camera operator's bodily movements and perceptions are recorded or 'embodied' along with the subject(s) in the frame".[19]

In addition to the way in which the physicality of the human body is bound up in cinematography, I argue that the thematizing of the camera operator also carries social meaning. Since human physicality is bound up with questions of gender, race, age, and so on, "cinematographic embodiment" also contains information about power dynamics at play while filming. This is an aspect that matters especially for women filmmakers and the representation of women's (traveling) bodies, because women have traditionally been looked at rather than been the bearer of the look.[20] The heritage of a long domination by male filmmakers as the exclusive holders of means of production to create filmic representations of women as well as the ongoing objectification and sexualization of women's bodies in entertainment and advertising today illustrates how the legacy of predominantly male camera-holding bodies has helped to create a system in which women have been the prevailing object of the gaze, with little or no authority over the representation of their bodies.[21]

Making visible the bodily and social dimensions of the camera-holding body can provide a way to challenge this legacy. Addressing the role camera-holding bodies play in creating filmic images can help to reveal that every image is carrying the footprint of a culturally and historically situated way of knowing and viewing, the absence or presence of an agreement between the camera operator and the object of the gaze, the incentive of economic gain, power relations, gender roles, expected ways of behaving, and so on. Locating cinematography in the body and making the camera-holding body perceivable allow the audience to see how filming is both a bodily and a social practice. It renders cinematography not just as something that is concerned with showing things, but also with making evident the socially and culturally situated ways of seeing.

When filming our documentary, for example, I often found myself in situations where I felt like the camera served as a shield we carried close to our bodies in order to protect us. In the streets of Cairo, for example, we found ourselves exposed to increased male attention and used our cameras as a buffer for looks we experienced as uncomfortable. The camera became a body that could return the gaze when our own bodies could not, it became a surrogate body, which we used to regain a sense of agency. At the same time, however, the camera also made us conscious of the way we were viewing others. The camera served as an eccentric mirror in which we recognized our own gaze, a fascination with everything that was unfamiliar to us. In "The Persistence of Vision", Donna Haraway addresses this possibility of vision to escape "binary oppositions".[22] While the "conquering gaze from nowhere [...] mythically inscribes all the marked bodies",[23] she highlights how we all depend on a body to see *our* body. Vision, in other words, is always embodied; seeing and knowing others and self is always situated.[24] Instead of enforcing the illusion of a seemingly neutral observation through the camera lens, embodied cinematography can help acknowledge and cultivate seeing as located in a situated body, and it can help us translate this notion of an embodied gaze to an audience.

Laura Marks, who explores film as a body to which viewers can relate to in a visceral way, offers insights into what happens during the process of presenting lived experiences via the filmic medium. The popular saying that certain films can *feel like* a punch in the stomach illustrates what is at stake here: although film cannot technically replicate the senses of touch, taste, and smell, Marks argues that film can appeal to and represent those senses, which are closely connected to the body.[25] Robin Curtis, for example, points out that film can emerge as a body on its own to which the viewer relates and responds in an intuitive way[26] and Vivian Sobchack explains that our vision can retain the experiences of the other senses, which makes it possible for us to haptically perceive without our bodies coming into direct contact with the material world:

> My sense of sight then is a modality of perception that is commutable to my other senses, and vice versa. My sight is never only sight – it sees what my ear can hear, my hand can touch, my nose can smell, and my tongue can taste. My entire bodily existence is implicated in my vision.[27]

The idea put forward here is that the sensation of embodied selfhood, which is based on impressions, can also be reproduced in film. Marks expands on this and argues that film can "evoke the particularly hard-to-present memories of people who move between cultures, by pointing beyond the limits of sight and sound".[28] This suggests that film can provide access to the ineffable, inexplicable, and often invisible elements of culture that are cultivated at the level of the body and learned through everyday interactions in specific cultural settings.[29]

One of the ways film can provide this access, according to Marks, is by means of haptic visuality, which she describes as a way of seeing that draws upon visceral experiences and multiple senses, a "kind of seeing that uses the eye like an organ of touch".[30] Working in dialogue with optic images which demand separation so that we can perceive the objects in them as distinct forms, haptic images are often unrecognizable and demand a more intuitive and embodied interpretation. An example of haptic visuality is when one looks at objects from extreme proximity so that the human eye might be able to sense the materiality of the object without discerning what exactly the object is. This means that haptic images are in a way less complete, leaving the viewer wondering what it is they are actually seeing.

For the representation of women's traveling bodies, the encounter with the other, and the depiction of foreign places, this bears the intriguing prom-ise of a visuality that is not easily consumable and therefore complicates instrumentalization. By denying clarity, haptic visuality then can obstruct, for example, a capitalist propensity to turn the female body into a projec-tion foil or a carrier of symbolic values with easily translatable meanings.[31] It can provide a possible strategy to counter instrumental and symbolic representation. By opting for a closer, less dominating, and less complete way of seeing, alternative strategies of knowing are activated. This is why Marks suggests that a focus on the haptic qualities of image production can "frustrate passive absorption of information"[32] and encourage viewers to engage more actively and self-critically with the image.

In our documentary, the concept of haptic visuality also provided us with a way to say the things we were unable to express with words. In our own film, we have come across situations and emotions that were difficult to grasp in words, let alone on camera. Given the fact that we were filming our journey as we were experiencing it, some of the problems of capturing situations or emotions have to do with the practical limitations of not always being able to film. At most border crossings, for example, we were not allowed to film. Other times, the difficulty of expressing lived experience stems from the lack of adequate language that was available to us at that point. An example which combines these two reasons, and which made us turn to haptic visuality, was the border crossing between Egypt and Israel where one of the border guards at immigration held on to Maria's passport and did not return it before making her acutely aware of the fact that he was turned on by her appearance. As we were not able to film this moment, the only record we have is one of Maria filling in migration and customs forms before we tried to cross with said police officer visible in the background. While we struggled to capture this experience adequately through words, we felt that it was important to create a filmic account that would allow access to the uneasy experience we lived through that day. After reading about Marks's theory of haptic visuality, I experimented with slowing down the image, zooming in and creating an unclear, blurry impression to represent that situation (see Figure 12.1).

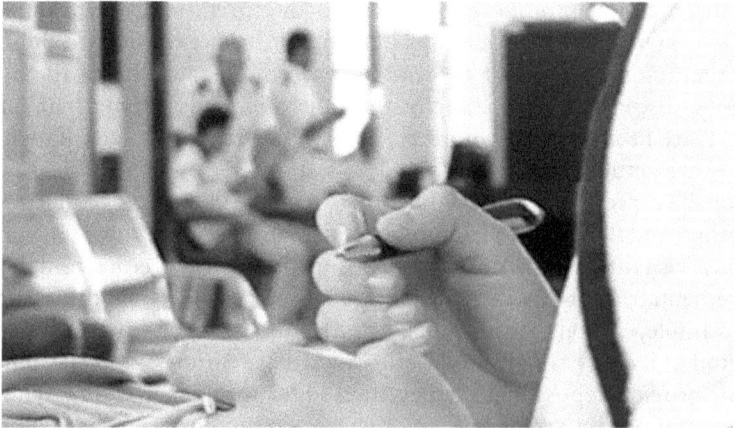

Figure 12.1 Film still from *Wanderlust*, showing Maria filling out the migration and customs form with the border guard in the background (Pérez Escalá and von Petersdorff, *Wanderlust* (at 0:13:30)).

The final scene documenting our border crossing between Egypt and Israel thus makes use of an interplay between optical images that situates the audience by means of establishing shots at the border and haptic images that seek to access our lived situation in a more visceral way. The blurredness of the image is meant to represent the unclear situation and a degree of disorientation that we felt being there, not sure how a reaction to this type of harassment might affect our plan of crossing the border.

As I argue for the importance of paying attention to the camera-holding body and the way images can affect a viewer, it is likewise important to acknowledge the role of the body in producing auditive elements of a film. Christine N. Brinckmann demonstrates how voice attains a particular quality from its origin in the body and a person's biography. She argues that an individual's voice has the unique power to transmit emotional states and to convey visceral presence because of its direct connection with the body.[33] Since the voice originates in the body, each individual's voice sounds different and depends not only on the shape and size of a person but also on the way this body and its speech have been formed.[34] Pronunciation, accents, pauses, intensity, and volume are all elements that are developed in an individual's life over time in a specific cultural and social context and convey information about a person's personality, upbringing, social class, and mood.[35]

When it comes to documentary film, it is noteworthy that the relationship between voice and body is often marked by an endeavor to keep the two separate. This becomes strikingly perceivable when looking at one particular form of voice-over in documentaries, often referred to as the 'Voice of God'. Predominantly found in classical documentaries of the 1930s and 1940s, the 'Voice of God' is characterized by an omnipresent, authoritative but disembodied voice that guides the viewer through the film and provides meaning to what is displayed on the visual level from a position of alleged superior knowledge. While the 'Voice of God' is per definition never embodied – the voice-generating body never appears visibly on the screen – it is of course not *bodiless* but traditionally white and male. In *The Acoustic Mirror: The Female Voice in Psychoanalysis and Cinema*, Kaja Silverman points out that female subjects are far more likely to be held to a unity of sound and image.[36] When a female voice materializes, Silverman argues, she is also expected to become visible as a body.[37] As a consequence, the female voice is often kept from being endowed with authority because "the voice-over is privileged to the degree that *it transcends the body*. Conversely, it loses power and authority with every corporeal encroachment, from a regional accent or idiosyncratic 'grain' to definitive localization in the image".[38] The disembodiment of the 'Voice of God' is possible, then, *because* it is uttered from a male body, while other voices are not because they are uttered from a female or marginalized body.

Finally, voice is not only something created in and through the body, but also *experienced* through the body. Similar to haptic visuality we might think of voice as having the ability to affect an audience in a visceral way. Brinckmann explains how voice can work as a "second track of sensual contact"[39] and highlights how it fulfills important psychological functions since it allows us to understand each other not only on a factual but also on an emotional level:

> Bekanntlich ist das Gehör der Sinn, der psychologisch am wichtigsten ist (auch wenn den Augen weit mehr praktische Bedeutung zukommt), weil er Wirklichkeitsdaten erfasse, die sich in unmittelbarer Nähe abspielen, im Rücken in der Dunkelheit, im Versteck. Es ist außerdem der Sinn, über den sich sprachliche Kommunikation vollzieht – und zwar nicht nur in ihrer verbalen Sinnvermittlung, sondern zugleich und auf einer tieferen Ebene als Vermittlung emotionaler Gegebenheiten.[40]

My own practical experiments with creating voice-over for our film have refined my understanding for the unparalleled ways in which voice can connect *what* is said with *how* something is said and how that links a narrating subject with a body that moves through a material world. As my body moved from Egypt to Germany, and later via the USA to Argentina to edit our film, my voice-over became a place of reflection of my lived experience in terms of *what* was said, but also through *how* something was said. We arrived in Germany in July 2014 but did not record our voice-over until the end of 2016 in Argentina. The temporal and spatial lag between the end of our journey and the recording of our voice-over materializes in two main ways: first, the temporal and geospatial changes manifest themselves in terms of how my voice changed from a more neutral Spanish accent I have in the on-camera comments in the film, to a much more Argentinian accent (with certain local attributes from the area of Buenos Aires). My intonation, pronunciation, and choice of words were no longer the same as they were when I set out on this trip with Maria. At the point of voice recording, I had been in Maria's hometown, La Plata, for several months, and as a result of that my Spanish had changed and I had adapted to the idiomatic expressions and intonation I heard on a daily basis. Secondly, and perhaps more significantly, the temporal lag becomes perceivable in terms of how memory and distance to the lived experience influenced my thoughts. The physical movement of my body to Argentina also had an influence on the way I reflected on our journey, as I now had a deeper understanding of the culture Maria grew up in and was able to relate better to some of her ideas and concerns she expressed throughout our journey over a year ago. The visceral difference in my voice and the way I was reliving parts of our journey with Maria while working on the film in post-production exemplify the transformations of my lived body through time and space. This demonstrates how embodied voice can exhibit the time gap between the 'I' that narrates and the 'I' that is experiencing a situation as well as the psychological development that the 'I' undergoes when translating between different cultures and languages.[41]

Feminist scholar Shirley Neuman points to the importance of a correspondence between a narrating 'I' and an experiencing 'I'.[42] She highlights how a conception of the autobiographical subject as a product of discourse must simultaneously include the autobiographical subject as a body that is "also a product of oppressive historical and material circumstances".[43] She argues that not recognizing how power relations are most fundamentally experienced at the level of the body denies women the experiential and corporeal sources of self-knowledge.[44] The following analysis of selected scenes from Beckerman's and Misselwitz's films provides examples of this and employs filmic strategies such as embodied female voice-over, embodied cinematography, and haptic visuality in order to foreground these "experiential and corporeal sources of self-knowledge".[45]

12.2.1 *Eine flüchtige Reise nach dem Orient*

In Ruth Beckermann's *Eine flüchtige Reise nach dem Orient* (1999) [*A Fleeting Passage to the Orient*], the Austrian filmmaker sets out to follow the footsteps of Empress Elizabeth of Austria (1837–98) on her journey to Egypt in the late nineteenth century. As a non-linear travel report, her documentary film takes the form of a self-reflective essay on foreignness and the way myths and exotic stereotypes are created by images. Beckermann interweaves her own travel experience with that of Empress Elizabeth, and while the film's primary focus is on the latter's journey, it is filled with meditative reflections and practical implications on Beckermann's filmmaking journey.

Drawing on the concept of embodied cinematography and embodied voice-over, Beckermann's *Reise* provides a powerful example of how a traveling body inscribes itself on the level of filmic image and sound, not through the camera operator's or director's bodily movements that are translated to the image, but through the reception and perception of the female filmmaker behind the camera. When it comes to representing the female body in *Reise*, it is striking that there are no actual images of either Elizabeth or Beckermann in the film. The one exception is the very beginning of the film when the opening scene shows a person (presumably Beckermann) on a train holding up a photograph of Elizabeth. In an attempt to interpret the absence of the female body in *Reise*, it is interesting to note that Beckermann recounts how Elizabeth refused to be photographed after her 31st birthday (Beckermann 0:02:00). Besides the common interpretation that Empress Elizabeth wanted to maintain a public image of eternal beauty, this refusal can be understood as a rejection of a means of representation that can be used to instrumentalize and foster stereotypes. Coinciding with the end of the nineteenth century, a time when photography became popular (and with it the notion that photographs accurately depict reality), Elizabeth's refusal to appear in photographs can be interpreted as an act of resistance: if women are objects to be looked at, the age of photography can be understood as exacerbating the mechanisms for realizing such a regime of seeing, while also diminishing women's autonomy over the representation of their bodies. Refusing to have her body photographed and thereby defying being marked that way can thus also be construed as part of a critical understanding of how visual representation works.

Despite the absence of the filmmaker's body in the images, *Reise* strongly thematizes the presence of the camera-holding bodies (the camera operators' bodies, Nurith Aviv and Sophie Cadet, and the body of the director, Ruth Beckermann) in creating the images. The camera often meets the gazes of people on the street, on markets, and in trains, calling attention to the bodies behind the camera. Especially throughout the first part of *Reise*, there are many moments where people look into the camera, wave at it, react to the female camera operators, or literally point out the camera in some other way. It is not so much the bodily movement of the camera-holding bodies which we perceive with the subjects in the frame, but the way people who appear on camera are affected by the bodily presence of the filmmaker behind the camera. As white, female, and Western, Beckermann's body is an 'exotic' visitor in Egypt and respectively causes attention. In her voice-over Beckermann is prompted to thematize the 'impact' of her appearing in Cairo: "Hier ist es unmöglich in der Menge zu verschwinden, es gibt kein Inkognito, alle zwei Minuten sagt jemand 'Hello how are you?' 'Where are you from?' [...]. Und geht neben einem her, ob ich nun antworte oder nicht macht keinen Unterschied" (0:21:00).[46]

Describing her own journey, Beckermann's voice-over thus links the narrative voice (who speaks) and focalization (who sees) with a body that has tangible, embodied experiences. While she refuses to tie her female voice-over to the visible appearance of her body, her voice-over reflections inscribe her body in the filmic medium and help to create an ambiguous representation of her travel experience. The traces of Beckermann's bodily presence in her film occasionally provoke (perhaps unwanted) involvement from the audience: every recorded image of a person looking at Beckermann's camera is not only an encounter between the filmed subject and the filmmaker but also an encounter with the audience of the film. The 'eye contact' established between her and a person she captured on the street in Egypt is postponed and dislodged in time and place to her audience.

As the people captured on camera look straight into the lens and reveal the recording device, the audience who looks at the screen is caught in the process of looking and prompted to reflect on the illusion that the camera captures people in a sincere and natural manner. The person looking back at Beckermann's camera thus renders the position of the audience as that of a distanced, safe observer as inoperative, and instead reveals the onlooker in the process of looking. This visual contact also reveals the filmmaker as a body that can be looked at and scrutinized and creates visual ambiguity. In traditional documentary filmmaking, the camera as a recording device typically demarcates a clear line between film*matter* and the film*maker*, the former referring to the object being filmed, the latter referring to the originator of the look. Since Beckermann becomes perceivable as a social actor, she does not escape representation. Centralizing this ambiguity in herself allows the audience to see Beckermann as both a subject and an object, opening up a more complex relation of power than a rigid film*maker* vs. film*matter* division does.

How to look without dominating and how to represent something or someone without mastering are then also among the central questions asked by Beckermann as she follows Empress Elizabeth's footsteps in Egypt and wonders how to resist the creation of stereotypical images, both of Elizabeth and of Egypt. In her voice-over, Beckermann wonders: "Sich das Bild von einem Menschen machen – ist das möglich? Hat es sie [Elizabeth] überhaupt gegeben? Oder ist sie eine Projektionsfläche unserer Träume und Wünsche und Fantasien, wie der Orient?" (Beckermann 1:04:36).[47] The interplay between the visual encounters with the people on the streets and Beckermann's own reflections about her journey encourages the audience to embark on their own inner exploration of what they "see" in the "Orient" and to revise stereotypical assumptions and images. Looking thus emerges as an intersubjective process in which there are no clear or stable binary roles. Instead, it is presented as a process that is marked by constant (re-)negotiation and fluidity.

12.2.2 *Tango Traum*

Unlike Beckermann's *Reise*, Helke Misselwitz's *Tango Traum* (1985) [*Tango Dream*] does not take the audience on a journey that physically transports the filmmaker's body to another place, but on an imaginary journey. Misselwitz's 20-minute film begins with an image of a ship on a frozen sea and contrasts this scenery of winter with non-diegetic tango music. As a starting point for Misselwitz's imaginary journey, the image alludes to the places and experiences that will remain inaccessible to her.

Foregrounding her own body, *Tango Traum* offers a distinctly different approach for capturing and representing women's embodied experiences than the one taken in *Reise*. The cinematography in *Tango Traum* can be described as static, nonspontaneous, scripted, and carefully constructed. Similar to still photography or paintings, the camera has a fixed field of vision with little or no motion. Long, calm shots provide the audience with time to analyze the individual elements and structure of the image and thereby encourage the viewer to acknowledge the materiality of what is being shown. Interspersing images of Misselwitz's body in close-up, repeatedly obscuring her face, and accentuating her bodily movements present the viewer with incomplete images and encourage a way of seeing that focuses on the haptic qualities of visceral sensations implicated in the images. Activities such as dressing, ironing, typing, smoking, and dancing are also portrayed in radical proximity, showing separate body parts so that the audience only gets a fragmented impression of a woman who is getting ready for a night out: a hand brushing over a black dress, a thigh as she puts on her stockings, an ear, and some hair as she puts in her earrings. The graininess of the footage, the close-up of her skin, and the fabric of her dress further invite the audience to draw on a tactile way of seeing.

Following Marks's insights about haptic images to access ineffable elements of a culture, Misselwitz's cinematography can be interpreted as an attempt to translate a sense of embodied selfhood to the filmic image. Through the activation of senses that are close to the body, such as touch, taste, feel, and smell, the cinematography in *Tango Traum* helps to anchor the audience in the materiality of Misselwitz's body and her apartment in the GDR (German Democratic Republic). The stationary, intimate aesthetic of the cinematography in *Tango Traum* is also representative of Misselwitz's sedentary situation as she physically remains in the limited space of her apartment for the duration of the film. Being bound to this physical place also marks the starting point for her voice-over narration:

> Buenos Aires und Montevideo sind weit weg, sehr weit weg. Von dieser Frau, die an ihrer Schreibmaschine sitzt und sich fragt, wie sie aus all dem einen Film machen könnte. Seit Wochen liest sie alles, was sie über Tango bekommen kann; hört Schallplatten, sieht sich alte Filme an [...]. Um zu empfinden, was Tango ist, legt die Frau immer wieder die Musik von damals auf. Ihr Herz nimmt die Musik an, aber Kopf und Herz treffen sich nicht. Alles liegt wie hinter Glas.
>
> (Misselwitz 0:01:49)[48]

As these words are uttered, the audience sees the filmmaker herself at her typewriter. While her hands and upper body are visible, her head remains visually obscured as if to exemplify the perceived disconnection between the head and heart that do not meet. Strikingly, the voice-over in *Tango Traum* expresses itself in the third person. While the idea suggests that it is most likely Misselwitz we are hearing, the source of the commentary remains undisclosed in the credits. Since the voice-over in *Tango Traum* expresses discontent with the inability to move around freely (and thus voicing an implicit critique towards the GDR government), a clear connection between the visible body of the filmmaker and first-person narrator could run the danger of censorship or even more severe consequences for the filmmaker.

Representing a materiality we cannot see but perceive through hearing, the voice-over illustrates how something that is out of sight and intangible fails to be simply absent. Another example of this can be found in a scene towards the end of the film where we see Misselwitz's naked back as she lies on her bed, her voice-over recounts:

> Seit Tagen hat die Frau immer wieder den gleichen Traum. Die feuchte Hand auf ihrem Rücken, die durch den Stoff ihres Kleides auf ihre Haut brennt. Der Druck der einzelnen Finger zeigt ihr an in welche Richtung sie ihre Füße setzen muss, der Druck der Schenkel dem sie zu entfliehen versucht. Der heiße Atem schlägt ihr ins Gesicht. Sie wendet den Kopf ab, er trifft ihren Hals, er erregt sie, sie fügt sich, für den Bruchteil einer Sekunde. Der Druck der Schenkel bringt sie wieder zur Besinnung. Sie gibt ihm erneut Schritte vor, denn es ist ja ein Spiel und eigentlich will sie es, immer wieder, bis zur Erschöpfung. Aber, warum verlässt er sie? Die Musiker verbeugen sich. Sie setzt sich auf ihren Stuhl, fächert sich Luft zu und wartet innerlich fiebernd auf den letzten Tanz.
>
> (*Tango Traum* 0:12:40)[49]

Her breathless narration and the stillness of her body create an incongruous sight. Her voice transmits a visceral sense of her excitement as she describes this imaginative tango dance. It is through the quality of her voice, the velocity, tonality, intonation, and timbre, that the audience gets a sense of her longing and the inability to still that longing. The voice-over slips free from material reality and glides into an in-between space, neither here nor there, neither in the GDR nor in Argentina and Uruguay. Articulating the distance as well as the connection to her lived, embodied reality, the voice-over provides a visceral access to the lack of contact Misselwitz makes perceivable through her film. Appealing to the haptic and the sensory, *Tango Traum* thus bears the marks of absence of an embodied encounter and provides an account of how the self is negotiated in absence of the other or in relation to an imagined place and time that is physically unattainable.

12.3   Conclusion

As the traveling body makes an appearance both visible and invisible, through embodied voice, in front of the camera and behind it, film gives access to ineffable and invisible elements of culture that are cultivated and expressed at the level of the body and in interaction between individuals. In the liminal space of unfamiliarity and encounters with other cultures, embodied filmmaking can cultivate spaces of uncertainty and fluidity. The act of translating one's body into new cultural and social contexts provokes an experience of self as unstable and open to negotiation (Beckermann's intersubjective camera is an example of that), and the act of travel – whether imagined or real – can also provide the traveler with a different perspective on familiar environments (Misselwitz's critique of the GDR can be seen as an example of this as it emerges out of her engagement with the dancing culture and history of Argentina and Uruguay).

Not being fixed to a place, an idea, or an image might then encourage the woman traveler to negotiate not only herself but also her relationship with those around her. In our own process of negotiating our traveling bodies in unfamiliar terrain, film has helped us to engage in a type of mediation between self and other in the moment of image production, but also as a temporal intermediary between the past, present, and future. As we edited the scene in our film which takes place in Aswan, Egypt, for example, it became apparent that I stuck out as a white, Western woman with a camera in my hand, but it was not until I saw footage of myself that I realized just how out of place my body was. In one of the first scenes in our film, Maria filmed me walking through a market in Aswan. As we edited this part and looked at the footage to record the voice-over, I was prompted to reflect on myself as an object, observing my strangeness, unfamiliarity, and foreignness. Entrapped in my own body, I am bound to a limited view, but with the help of the camera and my collaborator, Maria, I can look at myself from different angles. In that sense, the camera affords the ability to visually examine the self, the lived body as an external object to which thoughts can be directed.

Strategies of embodied filmmaking can thus be useful in exploring the negotiation happening between self and other in unfamiliar terrain. While the camera can, for example, serve as a shield in moments of vulnerability, embodied cinematography provides a multifaceted report of encounters with others and prompts the filmmaker to reflect on them. Through filmic strategies that account for this multilayered experience of moving our bodies through unfamiliar terrain, we might generate an ambiguity for the audience that opens the door to explore multiple meanings, new sides of self, contradictions, and discrepancies as well as one's cultural baggage and position in global power structures. Following the notion that meaning is based on one's position of reference to someone or something else, travel and movement – literally and figuratively speaking – can complicate the allocation of single meanings to subjects and places.

Notes

1 Maurice Merleau-Ponty, *Phenomenology of Perception,* trans. Colin Smith (London: Routledge, 2005), p. 409.
2 Merleau-Ponty, *Phenomenology*, p. 409.
3 Alain de Botton, *The Art of Travel* (London: Penguin, 2014), e-book, chapt. 2.7, loc. 547, par. 2.
4 Eric Leed, quoted in "Unraveling the Traveling Self", in Marguerite Helmers and Tilar Mazzeo, eds., *The Traveling and Writing Self* (Newcastle-upon-Tyne: Cambridge Scholars Publishing, 2007), pp. 1–19 (at p. 6).
5 Trịnh Thị Minh Hà, *Elsewhere, Within Here: Immigration, Refugeeism and the Boundary Event* (Abingdon: Routledge, 2010).
6 Alexandra Hidalgo, "Vanishing Fronteras: A Call for Documentary Filmmaking in Cultural Rhetorics (con la ayuda de Anzaldúa)", *Enculturation*, 21 (2016).
7 Jonathan Alexander and Jacqueline Rhodes, *On Multimodality: New Media in Composition Studies* (Champaign: National Council of Teachers of English, 2014).
8 bonnie lenore kyburz, *Cruel Auteurism: Affective Digital Mediations toward Film-Composition* (Boulder: University Press of Colorado, 2019).
9 Jody Shipka, *Toward a Composition Made Whole* (Pittsburgh: University of Pittsburgh Press, 2011).
10 Maria Pérez Escalá and Anne von Petersdorff, *Wanderlust, cuerpos en tránsito* [*Wanderlust, Female Bodies in Transit*]. Film. 2017.
11 As a bi-autobiographical account, the joint endeavor between Maria and I in making our documentary demands a clarification with regard to her contribution to the project: *Wanderlust* is a genuine collaboration on all levels of the filmmaking process (conceptualization, production, and postproduction).
12 Vanessa Veselka, "Green Screen: The Lack of Female Road Narratives and Why It Matters", *The American Reader*, 1, no. 4 (2013): n. pag., emphasis in text.
13 The countries we traveled through on our journey were: Egypt, Israel, Palestine, Southern Cyprus, Northern Cyprus, Turkey, Bulgaria, Serbia, Montenegro, Croatia, Bosnia-Herzegovina, Slovenia, Austria, and Germany.
14 Vivian Sobchack, *The Address of the Eye: A Phenomenology of Film Experience* (Princeton: Princeton University Press, 1992), p. 3.
15 Sobchack, *The Address of the Eye*, p. 3.
16 Ruth Beckermann, *Eine flüchtige Reise nach dem Orient* [*A Fleeting Passage to the Orient*]. Film. Filmladen, 1999. For better readability, the title will be shortened to *Reise*. Further references to this film are included parenthetically in the text.
17 Helke Misselwitz, *Tango Traum* [*Tango Dream*]. Film. Progress Film-Verleih, 1985. Further references to this film are included parenthetically in the text.
18 Michael Chanan, "The Role of History in the Individual: Working Notes for a Film", in Alisa Lebow, ed., *The Cinema of Me: Self and Subjectivity in First-Person Documentary Film* (New York City: Wallflower Press, 2012), e-book, sect. 4, loc. 648, par. 3.
19 Michael Albright, "The Visible Camera: Hand-Held Camera Movement and Cinematographic Embodiment in Autobiographical Documentary", *Spectator*, 31, no. 1 (2011): 34–40 (at p. 34).
20 See, for example, Laura Mulvey, "Visual Pleasure and Narrative Cinema", *Screen*, 16, no. 3 (1975): 6–18.
21 Based on Mulvey's influential essay in 1975 (see endnote 20), critics – and Mulvey herself – have since revisited the notion that a masculine subject position

is inescapable. Instead, Mulvey has assumed a more nuanced understanding of fluctuating viewing position which the (female) audience might assume. Despite these revisions and more refined understandings, women's 'to-be-looked-at-ness' and the sheer dominance of male protagonists continues to affect the way women and girls are perceived and represented through moving images.

22 Donna Haraway, "The Persistence of Vision", in Nicholas Mirzoeff, ed., *The Visual Culture Reader* (London: Routledge, 2002), pp. 677–84 (at p. 677).
23 Haraway, "The Persistence of Vision", p. 677.
24 Haraway, "The Persistence of Vision", p. 677.
25 Laura Marks, *The Skin of the Film: Intercultural Cinema, Embodiment, and the Senses* (Durham: Duke University Press, 2000), e-book, chapt. 3.1, loc. 2198, par. 1–2.
26 Robin Curtis, *Conscientious Viscerality* (Berlin: Dietrich Reimer Verlag, 2009).
27 Sobchack, *The Address of the Eye*, p. 78.
28 Marks, *The Skin of the Film*, chapt. 3.1, loc. 2203, par. 2.
29 Marks, *The Skin of the Film*, chapt. 3.2, loc. 2457, par. 14.
30 Marks, *The Skin of the Film*, chapt. 3.6, loc. 2732, par. 1.
31 Marks, *The Skin of the Film*, chapt. 3.2, loc. 2348, par. 4.
32 Marks, *The Skin of the Film*, chapt. 3.2, loc. 2276, par. 12.
33 Christine N. Brinckmann, *Die anthropomorphe Kamera und andere Schriften zur filmischen Narration* (Zürich: Chronos Verlag, 1997), p. 110.
34 Brinckmann, *Die anthropomorphe Kamera*, p. 110.
35 Brinckmann, *Die anthropomorphe Kamera*, p. 110.
36 Kaja Silverman, *The Acoustic Mirror: The Female Voice in Psychoanalysis and Cinema* (Bloomington: Indiana University Press, 1988).
37 Silverman, *The Acoustic Mirror*, p. 39.
38 Silverman, *The Acoustic Mirror*, p. 49, emphasis in text.
39 Brinckmann, *Die anthropomorphe Kamera*, p. 110.
40 Brinckmann, *Die anthropomorphe Kamera*, p. 110.
41 Swarnavel Eswaran Pillai, "The Texture of Interiority: Voiceover and Visuals", *Studies in Visual Arts and Communication: An International Journal*, 2, no. 1 (2015): 1–12 (at p. 4).
42 Shirley Neuman, "Autobiography: From Different Poetics to a Poetics of Difference", in Marlene Kadar, ed., *Essays on Life Writing: From Genre to Critical Practice* (Toronto: University of Toronto Press, 1992), pp. 213–30 (at p. 217).
43 Neuman, "Autobiography", p. 217.
44 Neuman, "Autobiography", p. 217.
45 Neuman, "Autobiography", p. 217.
46 "It is impossible here to disappear in the crowd, there is no incognito, every two minutes somebody says, "Hello how are you?" "Where are you from?" [...]. And walks next to you, whether I answer or not makes no difference." Unless otherwise indicated, translations of works are done by myself.
47 "Visualizing a human being – is that possible? Did she even exist? Or is she a projection screen for our dreams and desires and fantasies, like the Orient?"
48 "Buenos Aires and Montevideo are far away, very far away. From this woman sitting at her typewriter wondering how she could make a movie out of all this. For weeks she has been reading everything she can get her hands on about tango; listening to records, watching old films [...] In order to feel what tango is, the woman puts on the music from back then again and again. Her heart accepts the music, but head and heart do not meet. Everything seems to lie behind glass."

49 "For days the woman has had the same dream again and again. The damp hand on her back that burns through the fabric of her dress on her skin. The pressure of the individual fingers indicates in which direction she has to set her feet, the pressure of the thighs she is trying to escape. The hot breath hits her in the face. She turns her head away, it hits her neck, it excites her, she surrenders for a fraction of a second. The pressure of her thighs brings her back to her senses. She directs his moves again, because it is a game and she actually wants it, again and again, until she is exhausted. But why does he leave her? The musicians bow. She sits down on her chair, fans air and waits, feverish inside, for the last dance."

13 Tattooed Cartographies and the Displaced Body in an Age of Political Conflict

Karly Etz

13.1 Introduction

> We looked into an Armenian Convent, then visited the Tower of Hippicus, the only remaining one which Titus did not destroy, when he destroyed Jerusalem, & had a fine view [from] the top, then we walked back to our encampment through the Jaffa Gate. In the evening Dr. Rosen, who had so ably acted as our cicerone all the day, dined with us. I was afterwards tattooed by a native, & so was Keppel.[1]

This excerpt from Prince Albert Edward's journal describes a distinctive moment from his four-and-a-half month tour of the Middle East and the Mediterranean in 1862. Throughout his account, the prince discusses his trip as a pilgrimage of sorts, recording his presence at such sites of Judeo-Christian significance as the house where St. Peter lodged with Simon the Tanner, the ruins of the church where St. George was canonized, and the valley where Joshua's battle took place. Upon reaching Jerusalem, he is guided through the city by priests and taken to the holiest of these sites: Christ's tomb. Afterwards, the prince writes of his experience being tattooed by a Jerusalem native, along with his travel companion Keppel.[2] Albert's description of his time spent under the needle is minimal. Thankfully, a travel narrative written 19 years later by French journalist Gabriel Charmes describes the tattoo shop Albert visited on his journey in greater detail. In his account, Charmes notes that a certificate hung on the wall of the shop. The document boasted that the Prince of Wales had been satisfied with his tattoo of the Jerusalem Cross completed there on 2 April 1862 by tattoo artist Francis Souwan.[3]

The tattoo described in Charmes's account is a typical pilgrimage marking obtained by the Copts and other visitors to Jerusalem as a means of commemorating time spent in the Holy City, especially around Easter.[4] Today, travelers to Jerusalem still visit tattoo shops to obtain pilgrimage tattoos. Some of these shops have been under the management of the same family for centuries, passing down the trade from generation to generation. The process of tattooing pilgrims with souvenir designs has not changed

DOI: 10.4324/9781003331803-17

much in that time. Tattoo artists dealing in pilgrimage tattoos still use the same wood block stamps that their ancestors used hundreds of years ago, resulting in a general continuity of designs over the years. While many individuals continue to travel to Jerusalem with the sole intent of pilgrimage, Prince Albert's bodily souvenir was acquired as part of a much larger imperial program.

From the position of those in power, procuring tattoos was a typical component of travel vis-à-vis exploration and colonialism. European sailors often obtained tattoos both onboard ships and on land as a mark of their occupation, class, religion, and relationships and as proof of their successful arrival at a particular destination. Tattoos also traveled back to Europe on the bodies of Indigenous peoples. Prince Giolo, or "The Painted Prince", from Miangas Island was the first recorded tattooed person to be enslaved and brought to England from the Pacific in 1691. Giolo arrived in Europe at a time when explorers were increasingly scouring the globe for scientific and commercial purposes, bringing home fantastical descriptions of their travels that were immediately consumed by audiences hungry for information about exotic lands.[5] During the nineteenth century, tattooed ladies like Nora Hildebrandt published titillating travel narratives as proof of the shocking origins of their full-body coverings.[6] Many of these women claimed that Indigenous peoples captured them while they were passing through their territory and forced them to be tattooed as part of their captivity. However, upon return to their homeland, these women joined circus sideshows as a means to make a living, telling their stories to visitors who paid a small sum to gaze upon their mysterious bodily inscriptions.[7]

Prince Albert's story of acquiring a pilgrimage tattoo as an imperial souvenir serves as the historical antithesis to narratives of displacement discussed in the following essay. In contrast to the privileged position of a traveling prince surveying land under the watchful eyes of the crown, the contemporary artists creating tattooed cartographies over the past two decades have done so in an effort to draw attention to bodies forcibly displaced by moments of political rupture. As an itinerant medium grounded in lived experience, tattoos allow their wearers to visually map their travel through space and authenticate their experiences of displacement on those same traveling bodies. In this essay, I examine the work of three artists who build on the tattoo's historical relationship to bodily mobility and systems of power to formulate critical geographies. To do this, I first demonstrate the tattoo's ability to visualize travel through a discussion of the role of skin in perceiving and documenting the body's movement through space in the work of Qin Ga. I then connect this process to the activity of mapping in the work of Wafaa Bilal and discuss the ways in which the tattoo medium allows the wearer to perform his own traumatic experience of displacement and heal that trauma via the painful mode of tattoo application.

Finally, I conclude with Douglas Gordon's work as a speculative examination of the role of tattooed cartographies in addressing the impossibility of bodily travel across politically contentious border zones. In the following essay, the works of Qin, Bilal, and Gordon demonstrate the ways in which tattooed cartographies speak to disrupted notions of homeland and shifting conceptual boundaries by virtue of their location on the body's own boundary: skin. Each of the artworks examined here serves as a touchstone for discussions of the thematic relevance of tattooed maps in relation to the experiences of displaced subjects, namely their relationship to issues of mobility, marginality, nationalism, liminality, borderlines, (in-)visibility, unity, and division among others. In addition, these works expand the role of the tattoo beyond historically localized concerns and cultural traditions to a global medium ripe for twenty-first century artistic engagement – a shift that is especially poignant given the current tensions surrounding the immigration crisis and the rise of nationalism around the world.

13.2   Skin Documenting Travel

In order to understand the tattoo's ability to register travel on the body, we must first consider the role of skin in experiencing space. The surface of our body is what situates us in the world and defines the physical limits of our existence. French philosopher Maurice Merleau-Ponty, in his discussion of the body as a determinant of one's position or orientation within the world, emphasizes the body's relationship to space. He argues that without the body there is no perception of space, therefore the body is simultaneously in and of the space that surrounds it. In other words, "[t]o be located in space, which we all are, and to locate others, which we all do, requires embodiment".[8] Skin participates in our embodied perception of space through its delineation of the limits of the physical body and its primacy during activities of geographic perception. Geographer Paul Rodaway's study of "sensuous geographies" further articulates the body's (skin's) role in perceiving the world by defining four primary aspects of the body's surface that effect geographic perception: the body's *orientation* in the world, the body's ability to *measure* distance and scale, the body's *locomotion* and potential to explore, and the body's existence as a *coherent system* within which to coordinate all of its sensory experiences.[9] Each of these characteristics determines the body's perception of the outside world, designating the body's skin as a contested site, one that simultaneously embodies the limits of the physical self and establishes his or her place within the world. As Rodaway contests, "the tactile receptivity of the body, specifically the skin, [is] closely linked to the ability of the body to move through the environment [...]. [It] mediates between the body and the surrounding environment".[10] Thus, the skin's privileged position within phenomenological processes allows it to serve as a primary source document attesting to the lived experiences of bodies traveling through space.

A historical example of such documentation is visible in the tattooed skins of sailors returning from the South Seas during the eighteenth century. Upon their arrivals home, sailors used their tattoos as a means of substantiating their travel narratives and experiences. As Dorinda Outram asserts in her discussion of Enlightenment-era exploration, tattoos acquired by sailors during this time not only served as reminders of the vulnerability of the European body as it traveled through perilous waters but also served to enhance the moral economy of traveling bodies, deeming them authentic sources of knowledge.[11] Thus, tattooed skins fulfill an important role in the processes of travel narration. These markings establish the presence of the body within a specific geographic location and serve to substantiate the traveler's claims. Marguerite Helmers and Tilar J. Mazzeo discuss the importance of such confirmation in their discussion of the histories of travel narratives, stating "it is the promise of a physical body (and especially of a set of eyes) behind the narrative voice that claims our attention and interest. [...] [T]he conventions of the [travel] genre promise that its representations have their origins in the materiality of both the subject and the object".[12] As evidence of the body's movement through space, tattoos visualize the wearer's travel and serve as a method of authenticating lived experience.

Given the tattoo's ability to ground its owner's experience of space and place, it should come as no surprise that tattooed maps have become a familiar site among young people attempting to understand their position within the world today.[13] Jonathan Lewis addresses this phenomenon in his study of geographic imagery within contemporary tattoo communities, in which he posits that such markings "signify the wearer's consciousness of ties to the larger community in which the wearer is positioned".[14] Within the larger category of 'map tattoos', Lewis separates these markings into two main subcategories – those that fix identity within an officially bounded region and those that reject official borders in favor of depicting non-political geographic features or the Earth in its entirety. Those included in the former group allow their bearers to literally carry their home with them as they travel; this is why these types of tattoos are most popular among young people leaving home for the first time.[15] However, there is a clear distinction between the coming-of-age relocation represented by these kinds of popular map tattoos and the tumultuous moments of forced displacement relating to the artistic representations laid out here.

The lived body, or skin, is what affords us the knowledge of a particular space or landscape; however, the phenomenon of displacement derives from the failure of bodies to link up with places.[16] While general feelings of displacement represent the loss of particular places viewed as "home", the experience of 'forced displacement' specifically refers to "the often violent or destructive accumulated or sudden pressure that compels a person to make a desperate move away from their place of origin in pursuit of a place of hope and safety".[17] Several groups fall into the category of forcibly displaced bodies including asylum seekers, internally displaced persons, and documented and undocumented refugees.[18] These groups serve as the counterpoint to travelers who *seek* displacement as part of a journey to reach a desired destination.[19] The former faction is forcibly removed from their home-place and barred from return, while the latter faction voluntarily participates in displacement in order to facilitate adventure and relaxation.

As a means of locating, identifying, and bounding experience, tattooed cartographies allow artists to situate events, processes, and subjects within a coherent spatial frame and transform the experience of forced displacement into a format designed to move others and aid in their understanding of this phenomenon. This process of visualizing the travel of forcibly displaced peoples is realized in the work of Chinese artist Qin Ga, whose *Miniature Long March* (2002–05) recorded the artist's movement throughout China as he retraced the path of the retreat originally taken by Mao Zedong's Red Army between the years of 1934 and 1936.[20] When civil war broke out between the Nationalist and the Communist parties in 1927, battles raged throughout the country until a pivotal moment at the beginning of October 1934 when Nationalist forces encircled Communist headquarters in southwest China. The subsequent Communist retreat began as troops broke out of the encirclement at its weakest points, evading capture by Nationalist forces. Throughout the following year, the Red Army traversed approximately 4,000 miles and sustained thousands of casualties. The march's conclusion marked the emergence of Mao Zedong as the undisputed leader of the party, a result of his determination and endurance throughout his army's journey. Though the story of the Red Army's Long March is one fraught with catastrophe and turmoil, it has become synonymous with perseverance and prestige for the supporters of Mao Zedong's rise to power. In an attempt to critique this political narrative, Qin recreates and documents the lived reality of the Red Army's forced displacement through a cartographic tattoo on his own body.

Qin's *Miniature Long March* piece began as a means to document a much larger artistic collaboration titled *The Long March Project: A Walking Visual Display*, an undertaking that brought together Chinese and international artists, curators, and critics to visually engage with the utopian narrative of the Long March and its continuing relevance in contemporary Chinese society.[21] At its inception, the project's founder and chief curator Lu Jie established a 20-site journey following the Red Army's historical route along which participants set up a variety of relational artworks meant to engage with communities and local artistic practices. When the participants departed Beijing in June 2002, Qin stayed behind awaiting updates on the group's progression from site to site in order to record their movements on his tattooed back map. The artist's back had been tattooed with a basic map outline prior to the project's launch, including China's land border indicated by a purple line, provincial borders and their capitals marked with green lines and dots, another green line demarcating the Great Wall, and bodies of water including coastal borders marked in blue. As weeks passed, Qin denoted each newly visited site with a labeled red dot and connected it to its predecessor via a solid red line marking the path taken between them.[22] The creation of this tattooed national map created an arena upon which to delineate the experiences of individuals retracing the steps of displaced Chinese troops.

In September of the same year, the project came to a halt at Site 12, also known as Luding Bridge in the Sichuan Province. The combination of the grueling nature of the march and the remaining distance of the charted journey prompted the Long March team to declare the project an "incomplete completion", leaving Qin Ga's back map unfinished. Years later in May 2005, Qin decided to continue the group's march on his own, starting from the place they left off at Luding Bridge. As he traveled along remote roads and through mountain passes, he was accompanied by Gao Feng (a tattoo artist) and three cameramen, each of whom documented his journey in a variety of ways.[23] Though the first leg of the Long March Project involved Qin recording the movements of another group on his back, the second segment of the journey actively transcribed the artist's own body movements onto that same body. For Qin, this opportunity to enact the journey himself related directly to his foremost concern, representing the relationship between body and nature.

As he proceeded along the prescribed path, the mortality of his own flesh was undoubtedly at the forefront of his mind as he traversed sections of the journey abandoned by his collaborators. In an interview with Qin, he details the harsh nature of the journey and the perilous process of tattooing his body on site:

> Crossing the 5,000 metre high snowy mountains was probably the most difficult. The historical Red Army's experience during this period has become the subject of modern folklore. On our way up the mountain, it was raining and snowing. If we had stayed on the mountain for the night we wouldn't have had anything to eat, and our lighters didn't work. We got to the top of the mountain and the cameraman Li Ding couldn't go on. At that time the guide said that we had to go down immediately because Li Ding was very sick from the altitude, and that he might die if we did not immediately take him to a hospital. [...] He said that he would fight through, but for me to hurry up and tattoo.

He was throwing up as he waited for me. Gao Feng, the tattoo artist, is quite fit. He always carried the heaviest pack. But when he was tattooing, his hand wouldn't respond right.[24]

This narrative and Qin's tattooed map serve as evidence of the journey's toll on the body. The red additions to his back were not only indicative of the historical Red Army's name but also the bloody nature of its progress. With each stop, Qin's tattoo artist literally made his back bleed as he endured the addition of another point on the map and shouldered the burden of continuing the project himself. Though the artist's skin has healed, these red marks remind the viewer of the reality of his body's strenuous movement across the same country that decorates his back. The original Red Army was said to have traversed thousands of miles and sustained thousands of casualties. Through the reenactment of the Long March and the project's mirroring of bodies lost along the way, Qin stands as the lone soldier remaining, the toll of the journey documented on his skin. Here, the tattooed map and the artist himself serve as evidence of the harsh reality of a mythologized march. With each addition to his back map, Qin records his body's experience of the same march that displaced and killed thousands. As a result, his tattoo serves a similar function as the tattoos acquired by sailors visiting the South Seas. Each new line and dot upon his back visualizes Qin's movement through space and serves as a marker of authenticity, verifying his body's position throughout his travel across China, thus lending credibility to his personal account of the reality of the struggles associated with the historic Long March.

13.3   The Performative Pain of Tattooing

The tattoo's ability to authenticate lived experiences extends beyond the documentation of spaces traveled by the body and into the precarious position bodies inhabit through circumstances of forced displacement. By bringing the performance of tattooing into art spaces, artists like Wafaa Bilal transform the gallery from a comfort zone of artistic consumption into a conflict zone that exhibits the physical and emotional pain felt by populations displaced by war. Five years after the completion of Qin Ga's journey, Iraqi artist Wafaa Bilal created the piece *and Counting...* (2010) to document his experience of being a refugee and mourning lives lost during the Iraq War (Figure 13.1). Unlike Qin's descriptive map of China, Bilal presents a map primarily rendered in tattooed dots, without the inclusion of conspicuous demarcation lines. Prior to the date of his performance, Bilal had hired a tattoo artist to ink a borderless map of Iraq on his back. At that stage, the map solely consisted of the names of 16 Iraqi cities written in Arabic. No other geographic markers were present, although several of the cities that appeared in black across Bilal's back are located along the Tigris

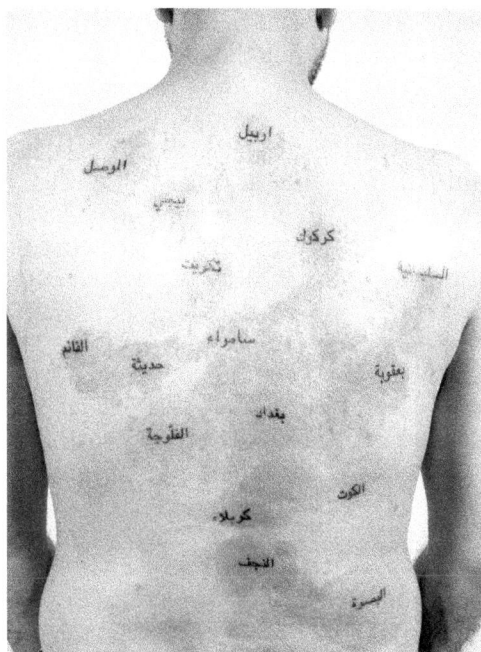

Figure 13.1 Photograph taken of Wafaa Bilal's back after being tattooed (Wafaa Bilal, "and Counting...", 2010. Photograph by Brad Farwell. Copyright: Wafaa Bilal. Published with support from the George Dewey and Mary J. Krumrine Endowment).

and Euphrates rivers. Without such conspicuous border lines, the audience is encouraged to focus on the body's own border – skin – and witness the painful process of tattooing as an empathetic witness rather than an outsider disconnected from Iraqi conflict by borders on a map.

On 9 March 2010, Bilal completed his tattoo map at the Elizabeth Foundation for the Arts in New York City. Over a period of 24 hours, the artist sat in a gallery as 105,000 dots were tattooed on his back. This number, according to Bilal, was the official estimated death toll in Iraq at the time of the performance.[25] A total of 100,000 dots, rendered using ultraviolet ink, represented the deaths of Iraqi citizens across the country, while 5,000 red dots recorded the number of American lives lost during the conflict. The pigments chosen for each set of dots reflects Bilal's sentiments regarding the visibility of the devastation in Iraq. While the dots rendered in red tattoo ink are visible to the unaided eye, those dots completed in ultraviolet ink require a black light to be seen.

Bilal's political statement, which criticizes the American media's emphasis on the deaths of soldiers over Iraqi citizens, is the product of his own experience of death back home and from afar as a forcibly displaced person. His adolescence and early adulthood were characterized by violence and political unrest in the Middle East. Growing up in the Holy City of Kufa, roughly 110 miles south of Baghdad, he experienced two Iraq wars and lived in constant fear, afraid of being labeled a dissident under Saddam Hussein's rule. In 1990, Bilal's fear became real when he was forced to run from governmental executioners after turning down Ba'ath recruiters who requested he join the war on Kuwait.[26] Bilal stayed in a Saudi Arabian refugee camp for the next two years, until 1992 when he was granted entrance into the United States.[27] Almost a decade later, he graduated with an MFA from the School of the Art Institute of Chicago, only to receive the news that his brother Haji had been killed by an American air-to-ground missile strike at a checkpoint on the outskirts of his hometown. Unable to return to his homeland, Bilal was forced to work through his anger and grief through his art.

Bilal's experience of the Iraq War and his desire to highlight unacknowledged Iraqi deaths at the hands of American soldiers led to the creation of his performance *and Counting...*, an artwork that simultaneously mapped the locations of lives lost while also exhibiting the bleeding Iraqi body. His choice of the dot to accomplish this task is poignant. In data mapping, points on a map designate an instance of certain phenomena, turning the death of a human being into a data point, symbolically distancing the viewer in some ways from the deceased. This visual marker mimics the distance orchestrated by the military in order to effectively attack the enemy with the same missile strikes that killed Bilal's brother. Today, many servicemen and women are now responsible for initiating attacks on foreign peoples from the safety of remote facilities, a practice that removes the body of the enemy from view a problem that Bilal rectifies in his work by forcing the viewer to confront the Iraqi body effected by such remote attacks. His attempt to bring a distant war to American soil recalls Chris Burden's seminal work *Shoot* (1971),[28] in which the artist had his friend shoot him in the arm as a way to bring the horrific realities of the Vietnam War to the American public. Each of these works highlights the suffering artist's body; however, Bilal's use of tattooed dots in *and Counting...* is one that simultaneously attempts to display the suffering body displaced by war and heal that suffering through the healing of the tattoo itself.

As medicinal tools, tattooed dots feature prominently in the traditional tattooing practices of rural Iraq, specifically within the peripheral nomadic tribes that border many Iraqi cities. These traditions were documented in an anthropological study completed by Winifred Smeaton in 1937[29] and continue within those groups today, although the practice is beginning to disappear.[30] Among these tribes, the practice of tattooing is still referred to as *daqq* or *dagg*, meaning "to strike" or "to knock" in Arabic.[31] While complex ornamental tattoo designs are still used to decorate women as a means of enhancing one's beauty and marking important moments of transition from childhood to adulthood, simpler geometric designs primarily serve therapeutic purposes such as curing ailments and relieving pain.[32] The latter type falls under the umbrella of what is referred to as "magical tattooing" of which Smeaton notes three varieties among Arab peoples: those designed to induce pregnancy, those designed to protect children against death, and tattoos used as "charms for love or against other magic".[33] Many of these tattoos, regardless of intent, were created using pigments derived from lampblack and sewing needles of varying numbers and sizes bound together and then coated in the mixture to be pricked through the skin.[34]

Though Arab tattooing traditions are largely geometric in nature, tattooed dots are particularly linked to healing. Within these nomadic tribes, tattooing one or several dots in targeted locations is often a method of insuring health among members. The most common type of healing tattoo is for sprains or headaches, or more general types of localized pain.[35] This kind of targeted application of tattoos to alleviate pain has existed for millenia in cultures around the globe. The oldest example of this practice can be found on the 5,300-year-old Tyrolean mummy known as "Ötzi" who was discovered in 1991 as climbers traversed the glacial region of the Ötzal Alps in Italy. Ötzi's skin contains groups of tattooed lines and crosses on areas of the body prone to overuse, such as joints and along the spine. The majority of these therapeutic tattoos correspond directly to Chinese acupuncture points, used to treat arthrosis and abdominal problems, leading many scholars to believe that his tattoos represent an early form of acupuncture.[36]

The healing qualities of tattoos are also visible in the work of Bilal. Though the process of rendering the piece *and Counting...* was a feat of endurance that made the artist bleed through the repetitive wounding of his skin, the use of ultraviolet tattoo ink meant that as soon as those tattoos healed they would seemingly disappear from the surface of his body. Unless made visible by a black light (Figure 13.2), each ultraviolet dot is invisible to the naked eye today (Figure 13.1). While Bilal's intention was to render visible the invisibility of Iraqi civilian deaths, the use of ultraviolet dots to represent lives lost allows each tattooed point on the map to perform the symbolic act of healing that loss. Just as the dots rendered on the bodies of nomadic Iraqi tribesmen and women serve medicinal purposes for their wearer, so too do Bilal's tattooed dots heal the trauma of forced displacement as a result of the Iraq War.

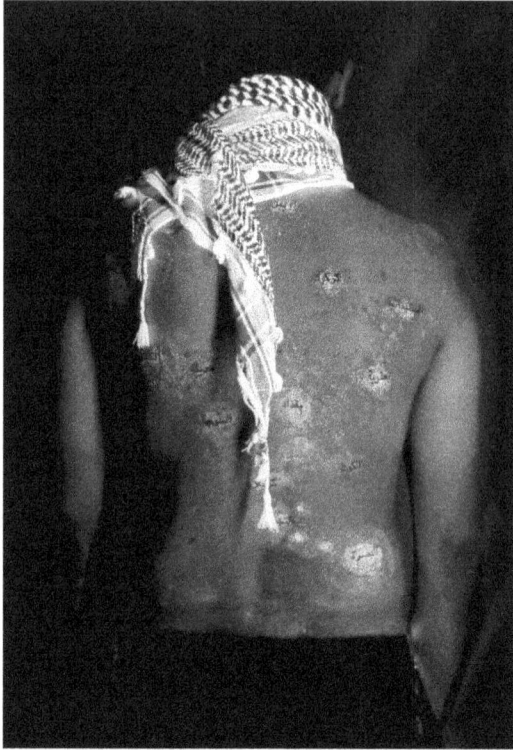

Figure 13.2 Photograph taken of Wafaa Bilal's back after being tattooed, under a black light (Wafaa Bilal, "and Counting...", 2010. Photograph by Brad Farwell. Copyright: Wafaa Bilal. Published with support from the George Dewey and Mary J. Krumrine Endowment).

13.4   The Impossibility of Travel

Wafaa Bilal's performance of *and Counting...* deliberately removes the contentious political borderlines that demarcate divisions between nation-states and feature prominently in experiences of forced displacement around the world. However, such demarcations are a reality and exist as significant markers of division serving to distinguish self from other and delineate one's home from foreign lands. As a final example, I examine an artwork that stands as a counterpoint to the forced displacement experienced by Mao Zedong's Red Army and Iraqi refugees and ask how do tattooed cartographies simultaneously stand as a metaphor for the body's displacement while also addressing the improbability of movement given a body's position within thickly bordered political states?[37] The seemingly impervious demarcation line featured in the work of Douglas Gordon exemplifies this experience of a sort of 'forced implacement' and seeks to rectify the Korean body's inability to traverse the contentious demilitarized zones (DMZ).

Gordon's film *Portrait of Janus (Divided States)* (2017) was completed as part of the UK/Korea 2017–18 Creative Futures season of cultural programming orchestrated by the British Council, a project celebrating positive cross-cultural interactions between the UK and Korea through artistic exchange and collaboration. The project's purpose comes to life in Gordon's film, which features tattoo artist Alexander Rasmussen tattooing the borderline between North and South Korea vertically along the spine of Korean man Janus Hoon Jung. As Rasmussen's needle renders the contentious demilitarized zone down Jung's back, South Korean cellist Okkyung Lee can be heard performing an original cello piece as an accompaniment to the buzzing of the tattoo needle. This performative collaboration echoes such avant-garde partnerships as Nam June Paik and Charlotte Moorman, who also addressed issues of cross-cultural contact between East and West in their work.[38]

Portrait of Janus begins with a black screen, allowing the viewer to focus on the sound of breathing and material crinkling. After about a minute, the source of the sound is revealed and we see the expanse of Jung's upper back wrapped in plastic. With every breath in, the plastic wrap clings tightly to his body and with every exhale the plastic releases, leaving indentations behind where the material was stretched taut across the skin. As the camera pans down, it gradually reveals a completed tattoo beneath the plastic. The covering of new tattoos with plastic is a common practice among reputable tattoo shops trying to prevent unintentional contamination or infection of the fresh wound. Once the entirety of Jung's back is shown, the film fades to black again and we hear the sound of Lee's cello entering in the background. She plays 16th notes in the key of E♭, continuing as Jung's back reappears, unmarked and *sans* plastic. The tattooist's needle is poised over a line drawn in blue ink, indicating the path it should take to correctly render the DMZ border. With every touch of the needle to skin, the buzzing of the tattoo machine joins Lee's 16th notes, accompanied by another sustained cello note in B♭. As the drone of the cello mixes with the buzz of the needle, the result is something akin to a Buddhist chant resonating in a monastery. Throughout the film, the tattoo is never shown in its entirety (Figure 13.3). Instead, Jung's skin consistently fills the frame generating the effect that the camera is capturing an aerial shot of land rather than a close-up of a tattoo on the skin, further connecting the subject's skin with geographic locations. This process continues until Rasmussen reaches the termination of the line and the shot fades to

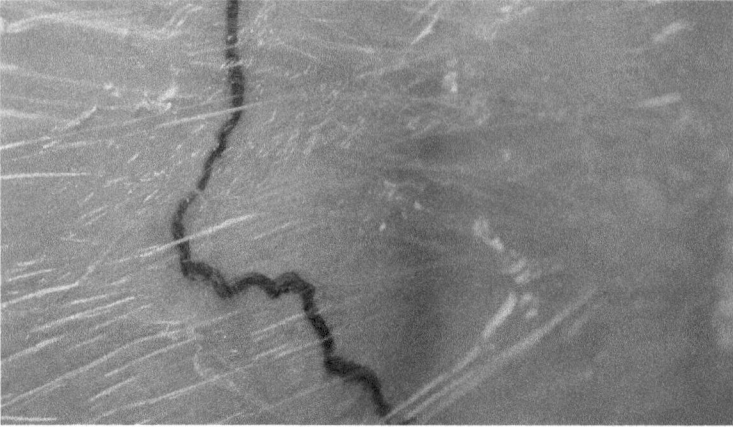

Figure 13.3 Video still from *Portrait of Janus* depicting a close-up shot of a tattooed line, with skin covered in plastic wrap (Douglas Gordon, *Portrait of Janus (Divided States)*, 2017 (still). Video, 1920x1080, 24 min. Copyright: Studio lost but found/VG Bild-Kunst, Bonn. Commissioned and produced by Locus+).

black again and the overlying accompaniment returns to the sound of Jung breathing and plastic wrap crinkling.

Gordon is not the first artist to emphasize the tattooed line in his work. Spanish artist Santiago Sierra featured the tattooed line for the first time in 1998 with the performance of *Person paid to have 30 cm line tattooed on them*[39] in Mexico City. For this piece, Sierra found a person without tattoos who was willing to have his back tattooed for 50 dollars, resulting in a permanent mark that would be carried around for life. Though Sierra has indicated that the use of the tattoo medium is irrelevant in contrast to the exchange of bodily autonomy for profit,[40] the use of the tattooed line inscribes the subject with a sign of marginality both in terms of his existence as a tattooed person in the late nineties and as a person literally on a borderline (given the line's presence on his body).

In contrast to the remuneration at the center of Sierra's performance, Gordon's work highlights the conflict surrounding Korea's DMZ. This political boundary has been a source of international tension since its establishment along the 38th parallel in 1953. While it continues to fuel conflict around the world, many contemporary Korean maps omit the line, choosing instead to represent the Korean peninsula as a united nation with Seoul as the capital. In place of a boundary line, a 'demarcation line' symbol marks the separation, a notation that lacks the divisive weight of the northern boundary with China.[41] Gordon's film highlights this often-elided borderline allowing the DMZ line to take center stage without the distraction of additional cartographic markers. However, his decision to rotate this line approximately 90 degrees to the right not only illustrates the division of the Korean population through the visual division of Jung's body, it also revises this division in an effort to allow bodies' mobility throughout the Korean peninsula.

We can better understand the tattoo's ability to counteract the impossibility of travel across borderlines by examining the intersection between the body's experience of space and cartographic borders lines. In Rosalyn Diprose and Robyn Ferrell's introduction to their philosophical treatise *Cartographies: Poststructuralism and the Mapping of Bodies and Spaces*, both authors establish a relationship between cartographic borderlines and the phenomenological body, stating:

> In the metaphor of cartography, to draw a line is to produce a space, and the production of the space effects the line. The map describes a territory as the compass describes an arc; the lines on the map produce borders beyond which things are seen to be different. Yet the difference of the 'outside' also defines what is 'inside' the border.[42]

This discussion mirrors spatial theory scholarship that places the body at the center of mapping processes, with skin serving as the dividing line between self and other. Both Diprose and Ferrell solidify this connection later on in their text, asserting that, "[a]ttempting to locate the body definitively, or, taking up a position by evoking one's own specificity, necessarily involves reference to an 'outside'".[43]

Relating human skins to borders on a map is an apt comparison given the skin's ever-changing surface and relative permeability. Though the largest organ on our body serves as protection from our surroundings, it is not infallible. As we age, the outlines of our body shift and wrinkle, changing as our skins record the passage of time and our interactions with the world around us. Cuts, nicks, burns, tattoos, and piercings document our histories and attest to the skin's fragility, as well as its inevitable transformation. Contemporary approaches to border studies position cartographic lines as similarly mutable constructs. For example, geographer David Newman's theoretical framework surrounding borders and power contends that borderlines should be viewed as flexible since they are reflective of constantly evolving patterns of human behavior.[44] This argument expresses the role of borders in the broadest sense, as a process of identity creation defined by the separation between "self" and "other". As the objectives and character of a group evolve over time, so too do borders shift to better define that group's relationship to the outside world.

Gordon acknowledges the potential for tattooed skins to signify mutable borderlines through his reorientation of the DMZ from a horizontal division between north and south to a vertical line which actively marries these currently divided states. This revision references a time in which a prominent cartographic line united the Korean peninsula rather than dividing it, allowing for the travel of bodies throughout the region. Prior to the creation of the DMZ, Korean mapmaking traditions reflected domestic values and foreign influence. John Rennie Short details this evolution in his book *Korea: A Cartographic History*, in which he examines several pivotal influences (both foreign and domestic) that affected Korea's cartographic trajectory. One style of Korean map that continued to influence mapmaking traditions for centuries in spite of these influences is the *hyongsedo* or "shapes-and-forces" style map that first appeared in the late fourteenth century.[45] This style of mapping derived from shamanistic and geomantic belief systems, including the notion that the land is a living breathing body with geographic features such as rivers conceptualized as veins containing the country's lifeblood.[46] The 1861 *Daedong yeojido* or "Map of the Great East [Korea]"[47] by cartographer Jeong-ho Kim (1804–66) is a prime example of the continued influence of the *hyongsedo* style. Short describes the map's employment of

> mountains as a unitary skeletal frame and rivers like blood vessels; the land is presented as a living body, with political and physical geographies combined into the national coherence of a living, breathing entity. Mountains are shown as continuously linked, sinuous sawtoothlines, the entire mountain chain system radiating out from Baekdu Mountain and back, suggesting both the underlying geomantic forces and the national cohesion.[48]

Other Korean maps completed during this period contained postscripts in which these geomantic characteristics were explicitly outlined, such as the *[D]aedong ch'ongdo*. Its postscript declares that

> [l]ooking at the topography of our land [...], [i]f Mt. [Baekdu] is the head, Taeryong (*Paektu taegan*) is the spine. The land looks like a man with a crooked back who stands with his head turned sideways, and Tsushima of Yongnam and Tamna (Cheju Island) of Honam are like two legs.[49]

Such postscripts attest to the fact that Koreans traditionally regarded their land as a human body, with the Baekdu mountain range functioning as the country's spine, connecting north to south in one continuous vertical line.

Through the reorientation of the DMZ border from a horizontal to a vertical line placed along Jung's actual spine, Gordon's film references this perception of the peninsula-as-body found throughout Korea's cartographic history. This perspective shifts our reading of Gordon's border from a representation of division to an indication of connectivity and national unity. With this repositioning, Gordon's collaboration with Jung and Lee rewrites Korea's divisive present by referencing its cartographic past. Furthermore, by tattooing the contentious DMZ border on Jung's mobile body, Gordon doubles down on his assertion that the Korean border is a mutable construct. As the line on Jung's back moves with him, it serves as a constant reminder of the mutability of borderlines and performs a speculative future in which the border between North and South Korea is altered to allow for the travel of bodies throughout the peninsula.

13.5 Conclusion

In the wake of the global immigration crisis and the rise of nationalism around the world, conversations surrounding the experiences of people forcibly displaced through moments of political conflict will continue to feature prominently within contemporary artistic practice. While maps continue to act as ready metaphors for the experience of displacement and processes of locating the body in space, tattooed maps maintain the unique ability to embody moments of displacement and authenticate the lived experiences of bodies traveling to escape political turmoil within a given region. The works produced by the artists discussed in this essay specifically utilize tattooed cartographies as reminders of the borders crossed by these itinerant bodies and the theoretical connections between processes of locating the body in space and mapping diverse experiences of displacement. The critical geographies of Qin, Bilal, and Gordon reject historical notions of authority attached to more normative maps and demand that the viewer reassess their understanding of contentious regions around the world. Highlighting the body's own border between self and environment, these tattooed skins render cartographies permanent upon the skin while simultaneously connecting maps to ephemeral bodies. These seemingly contradictory notions work to correct problematic perceptions of permanence tied to contentious cartographies while also authenticating the lived experience of individuals whose forced displacement and longing to return home are contingent upon the existence of these bounded regions.

Notes

1 Prince Albert Edward, *Cairo to Constantinople: The Prince of Wales Journal, February 6 – June 14, 1862*, Royal Collection Trust, pp. 33–35.
2 Edward, *Cairo to Constantinople*, p. 35.
3 Gabriel Charmes, *Voyage en Syrie*, vol. 45, pt. 2, June 1881, p. 771.
4 John Carswell, *Coptic Tattoo Designs* (Beirut: The American University of Beirut, 1958), p. xi.
5 Michelle Hetherington, *Cook & Omai: The Cult of the South Seas* (Parks: National Library of Australia, 2001), p. 37.

6 Jennifer Putzi, *Identifying Marks: Race, Gender, and the Marked Body in Nineteenth-Century America* (Athens, GA: The University of Georgia Press, 2006), p. 13.

7 Margo DeMello, *Bodies of Inscription: A Cultural History of the Modern Tattoo Community* (Durham, NC: Duke University Press, 2000), p. 58.

8 Rosalyn Diprose, "A 'Genetics' that Makes Sense", in Rosalyn Diprose and Robyn Ferrell, eds., *Cartographies: Poststructuralism and the Mapping of Bodies and Spaces* (Sydney: Allen & Unwin, 1991), pp. 65–76 (at p. 65).

9 Paul Rodaway, *Sensuous Geographies: Body, Sense, and Place* (London: Routledge, 1994), p. 31.

10 Rodaway, *Sensuous Geographies*, pp. 41–42.

11 Dorinda Outram, "On Being Perseus: New Knowledge, Dislocation, and Enlightenment Exploration", in David N. Livingstone and Charles W.J. Withers, eds., *Geography and Enlightenment* (Chicago, IL: The University of Chicago Press, 1999), pp. 281–94 (at p. 292).

12 Marguerite Helmers, and Tilar J. Mazzeo, "Introduction: Travel and the Body", *Journal of Narrative Theory*, 35, no. 3 (Fall 2005): 267–76 (at pp. 267–68).

13 Jonathan Lewis, "Tattoo-Communities and Map Tattoos", in Kori Duin Kelly, ed., *Bodily Inscriptions: Interdisciplinary Explorations into Embodiment* (Cambridge: Cambridge Scholars Publishing, 2008), 55–65 (at p. 56).

14 Lewis, "Tattoo-Communities", p. 56.

15 Lewis, "Tattoo-Communities", p. 61.

16 Edward S. Casey, *Getting Back into Place: Toward a Renewed Understanding of the Place-World* (Bloomington, IN: Indiana University Press, 1993), p. xiv.

17 Paul N. Sydnor, "Understanding the Forced Displacement of Refugees in Terms of the Person", *Transformation*, 28, no. 1 (2011): 51–61 (at p. 52).

18 Sydnor, "Understanding the Forced Displacement of Refugees", pp. 53–54.

19 Casey, *Getting Back into Place*, p. 308.

20 For a photograph of the piece see Qin Ga, "The Miniature Long March", QAGOMA, https://collection.qagoma.qld.gov.au/stories/353 [21 September 2022].

21 Timothy Shea, "Where There Are No Art Circles: The Long March Project and New Geographies of Contemporary Chinese Art", *Yishu* 16, no. 6 (2017): 77–83 (at p. 77).

22 Shea, "The Long March Project", p. 79.

23 Lu Jie, "Qin Ga: 'Miniature Long March'", *Artlink*, 25, no. 3 (2005): 52–56 (at p. 52).

24 Jie, "Qin Ga", p. 54.

25 This number has been highly disputed since the time of the Iraq War. "105,000" is a number that comes from Bilal himself and contradicts several surveys conducted both during and after the war. No study has been able to ascertain the exact number of casualties sustained over the course of the war although controversial studies have approximated such diverse numbers ranging from a conservative 150,000 deaths to over a million.

26 Wafaa Bilal and Kari Lyderson, *Shoot an Iraqi: Art, Life and Resistance Under the Gun* (Francisco, CA: City Lights, 2008), p. 68.

27 Bilal and Lyderson, *Shoot and Iraqi*, p. 141.

28 For more on Burden's oeuvre see Fred Hoffman, ed., *Chris Burden* (Newcastle upon Tyne: Locus+ Publishing, 2005).

29 Winifred Smeaton, "Tattooing among the Arabs of Iraq," *American Anthropologist*, 39, no. 1 (1937): 53–61.

30 Margo DeMello, *Inked: Tattoos and Body Art Around the World* (Santa Barbara, CA: ABC-CLIO, 2014), p. 314.

31 Smeaton, "Tattooing among the Arabs", p. 53.
32 Lars Krutak, *The Tattooing Arts of Tribal Women* (London: Bennet & Bloom, 2007), p. 36.
33 Smeaton, "Tattooing among the Arabs", p. 54.
34 Smeaton, "Tattooing among the Arabs", p. 59.
35 Smeaton, "Tattooing among the Arabs", p. 54.
36 M. A. Pabst et al., "The Tattoos of the Tyrolean Iceman: A Light Microscopical, Ultrastructural and Element Analytical Study", *Journal of Archaeological Science*, 26 (2009): 2335–41 (at p. 2338).
37 The concept of a "thick border" is one that Filippo Dionigi discusses in his article on the Syrian refugee crisis. In it, he posits a difference between 'thin' and 'thick borders' with the former maintaining scant containment capacity while the latter maintains separation of spaces and establishes a strict discontinuity between people and things contained inside and outside the borders ("The Syrian Refugee Crisis in the Kurdish Region of Iraq: Explaining the Role of Borders in Situations of Forced Displacement", *International Migration*, 5, no. 2 (2018): 10–31).
38 Perhaps the most compelling comparison can be made with Paik and Moorman's 1965 performance of John Cage's *26'1.499" for a String Player*, in which Moorman stretched a single cello string vertically across Paik's naked torso as he knelt between her legs. Moorman then proceeded to "play" Paik as if he was her instrument, a process that included regularly slapping and thumping him across the back. Paik stated later on that this sort of abuse was intended to represent America's cruel imperialism in Asia. For more information on Paik and Moorman's collaboration see Joan Rothfus, *Topless Cellist: The Improbable Life of Charlotte Moorman* (Cambridge, MA: The MIT Press, 2014), p. 113.
39 For more on Sierra's oeuvre see Santiago Sierra, Fabio Cavallucci, and Carlos Jiménez, *Santiago Sierra* (Milan: Silvana, 2005).
40 Santiago Sierra and Eckhard Schneider, eds., *Santiago Sierra: 300 Tons and Previous Works* (Köln: König, 2004), p. 27.
41 John Rennie Short, *Korea: A Cartographic History* (Chicago, IL: The University of Chicago Press, 2012), p. 137.
42 Rosalyn Diprose and Robyn Ferrell, "Introduction", in Rosalyn Diprose and Robyn Ferrell, eds., *Cartographies: Poststructuralism and the Mapping of Bodies and Spaces* (Sydney: Allen & Unwin, 1991), pp. viii–xi (at p. ix).
43 Diprose and Ferrell, *Cartographies*, p. 76.
44 David Newman, "On Borders and Power: A Theoretical Framework", *Journal of Borderlands Studies*, 18, no. 1 (2003): 13–25 (at p. 13).
45 Short, *Korea*, p. 26.
46 Short, *Korea*, p. 27.
47 For a photograph of the piece see Kim Jeong-ho, "Daedong yeojido", UWM Libraries, https://collections.lib.uwm.edu/digital/collection/agdm/id/829/ [4 November 2022].
48 Short, *Korea*, p. 98.
49 Han Young-woo, Ahn Hwi-Joon, and Bae Wood Sung, *The Artistry of Early Korean Cartography* (Larkspur, CA: Tamal Vista Publications, 1999), p. 8.

14 Strolling through the City on a Self-Guided Tour

Embodied Engagement with the Urban Space

Nora Winsky

14.1 Introduction

Nowadays, those privileged to travel the world frequently use their phones at all stages of their journey: reading travel blogs, booking tourist services on the move, posting recommendations in online portals, or following a displayed route within a 'new' city. Such examples show the varied implementation of mobile media in travel contexts. Travelers relate to both online and offline worlds. Jennie Germann Molz uses the term "blended geographies" for "hybrid assemblages of physical and virtual environments in which bodies, technologies, virtualities and materialities become entangled with one another".[1]

The tourism industry creates products that respond to the trend toward increased use of mobile devices.[2] Self-guided tours were invented as an alternative to traditional sightseeing tours and as a reaction to the growing number of digital applications. In the light of this trend, this chapter focuses on such a tour, which can be regarded as an application-driven sightseeing tour for visitors who want to discover urban narratives independently by following route maps through a historic city center. In Freiburg, a medieval city in Southwest Germany, users of an app by Freiburg Living History can listen to stories of the city's history and can use "augmented reality"[3] technologies to see historic views. This app, which was designed for visitors, serves as a case example in the following study.

My perspective is that material and immaterial components co-constitute each other and influence the bodily experience when strolling through the city on a self-guided tour. In this respect, the interactions between traveling bodies, urban settings, and mobile media are central and it is the interplay between these elements that affects the way in which the body engages with urban spaces. The case study exemplifies how the body gives attention to specific sites and their characteristics in the embodied act of a self-guided tour. Consequently, my analysis of bodily travel experiences based on media use is organized as follows: first, phenomenological approaches within human geography are introduced and the concept of the "sensory-inscribed" body[4] by Jason Farman is outlined; secondly,

DOI: 10.4324/9781003331803-18

the self-guided tour offered by Freiburg Living History is presented and qualitative research methods are applied to analyze the perceptions and experiences of traveling bodies. Participatory observations and in-depth interviews with the app inventors are the main source of the study. The findings concerning the mobile app are conceptualized and discussed with reference to Farman's theoretical considerations. Finally, possible future research aspects on multimedia experiences in tourism contexts are discussed.[5]

14.2   Phenomenological Approaches within Human Geography

Throughout the twentieth century, scholars of tourism preferred the visual[6] to embodied and multisensory tourist performances.[7] With the 'performative turn' in the humanities and the social sciences, in particular since the 1990s, tourist places were understood as sites of embodied practices, focusing on performances and embodiment.[8] A shift from "text and representation to performance and practice"[9] occurred. The continuous integration of phenomenological approaches within human geography emphasized the relevance of human agency, body, and embodiment.[10] According to Solène Prince, "[t]ourism scholars have challenged the hegemony of vision [...] in their conceptualization of tourism, making way for a more subjective, relational and negotiated notion of tourist landscapes and spaces".[11] In consequence, tourists as traveling bodies do not only visually consume unfamiliar places: they encounter new landscapes in the first person. In cities, tourists immerse themselves in urban everyday life. They discover unknown scents and flavors, as well as novel colors within the urban space. The multisensory perception, practices, and performances of situated subjects are highly significant in contexts of traveling.[12] The investigation of life-world experiences in subject-centered contacts with the 'new' and 'unknown' is an important research field located at the intersection of tourism studies, human geography, and phenomenology. The latter, in particular, emphasizes subjectively perceived and sensory experiences, describing and interpreting human experiences.

The body is a medium of ongoing involvement in the world. Therefore, according to Martin Heidegger, a traveling body is an embodied subject or "being-in-the-world" [Dasein].[13] Following this perspective, place-based research is interested in "a more corporeal [leiblich] experience in the life-world".[14] In this frame, one's own perceptions form the basis for experiences that are intuitive and spatial by nature. Here, bodily space is distinguishable from objective space as a measurable construction. As underscored in the previous paragraph, human perception (of space) goes beyond the five traditionally recognized senses (visual, auditory, gustatory, olfactory, and somatosensory). The senses do not act independently of one another; rather, the entire body experiences an overall feeling of places:

"As places and spaces are dimensional arrangements of the life-world, they are characterized by atmospheres, of which the subject – and only the subject – can give evidence (the 'width' or 'narrowness' of spaces, 'pleasant' or 'uncomfortable' places etc.)."[15]

The somatic body steadily experiences spaces as 'being-in-the-world' and is always surrounded by spaces and material objects (for example, streets, buildings, vehicles, and smaller artifacts such as garbage cans, signs, and posters). As David Seamon argues: "Every moment of our lives, we always find ourselves caught up and immersed in a world that is there before us in inescapable presence."[16] Hence, sensory perceptions and experiences are embodied activities. As enactive approaches claim, bodily skills and sensorimotor knowledge are constitutive of perceptual sensations that one understands.[17] Thus, practical knowledge is the basis for human beings to reflect and give meaning. In tourism contexts, traveling bodies interact as agents with unfamiliar environments. Such forms of interactions are also facilitated by technologies that extend physical capacities. Analyzing what traveling bodies do by using media devices while strolling through the city on a self-guided tour provides insight into how bodies and sensory perceptions, practical knowledge, and environments are linked. The sensing of the world is partially affected by technological mediation. Consequently, embodiment is a central element for analyzing the use of self-guided tours from a phenomenological perspective. Taking this into account, the following sections focus on the concept of embodiment and the "sensory-inscribed" body[18] as Farman discusses it.

The body is "a physical and biological entity, lived experience, and a center of agency, a location for speaking and acting on the world".[19] In general, embodiment is the body's expression of being in and acting on the world. In social sciences, different disciplines (including anthropology, cultural studies, and human geography) operate with the concept of embodiment, although they offer no distinct definition. As Shaun Gallagher remarks, these scientific fields consider embodiment as an integrative view of body and mind.[20] Thus, in every social situation, the interplay of corporeal and reflective perception organizes interpretation and agency. Thomas J. Csordas characterizes embodiment as an "indeterminate methodological field defined by perceptual experience and mode of presence and engagement in the world".[21] It is crucial that the world in its spatial extension always surrounds the perceptive subject. Ted Toadvine summarizes these aspects as follows: "The body's relationship with space is therefore intentional, although as an 'I can' rather than an 'I think'; bodily space is a multi-layered manner of relating to things, so that the body is not 'in' space but lives or inhabits it."[22]

Embodiment, in a tourist context, is a way of inhabiting the world and comprises sensory and corporeal engagements with specific sights, attractions, and monuments. Sightseeing activities go far beyond visual consumption and involve various embodied practices. Especially when away from home, engagement with the 'new' and 'unknown' appears as a broad field that the traveling bodies themselves are strongly aware of. Strolling through streets, standing in front of impressive monuments, trying out new dishes, observing daily life, or enjoying the wind on one's skin exemplify those embodied experiences. In foreign surroundings, it is much easier for people to pay attention to their interactions with and practices in the physical world. In sum, the awareness of the self as a traveling body is intensified.

Building on these considerations, Farman developed the conception of the "sensory-inscribed" body.[23] Here, he combines two approaches that function as a model for interactions with mobile interfaces: he draws on Maurice Merleau-Ponty's *Phenomenology of Perception*[24], and additionally on Jacques Derrida's poststructuralist approaches illustrated in *Of Grammatology*.[25] Farman understands bodies as "simultaneously conceived through site-specific sensory engagement and a reading of bodies as always culturally inscribed".[26] Conflating these two modes of embodiment – the body as a sensory system and as a sign system – allows us to describe and to study the production of embodied space in a digital culture – a paradigm shift that grants access to online information at any time and everywhere.[27]

Adapting Farman's model to tourist contexts integrates the symbolic dimension of reading processes in unknown surroundings: "Being-a-tourist-in-the-world entails constantly figuring and refiguring [...] identities, knowledge, values and meanings."[28] By drawing on their own cultural experiences and foreknowledge, tourists try to make sense of what they encounter. Culturally acquired knowledge and conduct – such as knowing how a sightseeing tour proceeds and how tour guides and participants take up different roles – is part of the embodied encounter. "[E]mbodiment can never exist outside of culture. Culture frames all our embodied and spatial interactions."[29] In the first place, sense perception and experience "depend on perceptual knowledge and on the skill with which you bring this knowledge to bear on what you encounter".[30] Beyond that, we grow up in specific cultural and social contexts that complement and extend our perceptual experience. These impacts influence the ways we read the world and therefore work in tandem with active perception.

A key aspect of Farman's concept is the way he incorporates mobile media and the body: they influence and constitute each other in inseparable online and offline worlds. In the following sections, a self-guided tour that can be used in the city center of Freiburg serves to describe the nexus of traveling bodies, mobile media, and urban spaces. On that account, qualitative research methods are applied.

14.3   Self-Guided Tour: Traveling Bodies Engaging with Urban Spaces

Since the new millennium, the growth of locative media has been driven by the spread of ubiquitous computing and geographic information systems, from GPS to Google Maps.[31] As tourists try to find their way through cities and orient themselves in unknown surroundings, it is not surprising that the tourism industry has taken advantage of these technological developments. Online travel guides present different attractions on a map, complemented by helpful information. Furthermore, travel communities like TripAdvisor provide users with reviews of restaurants that are located in geographical proximity to their own position. These new possibilities of obtaining geographical information and wayfinding also influence the way a city is perceived. In addition, GPS tracking systems continuously provide information about the user's current position and requested location, providing real-time information about their surroundings. As Jen Southern puts it: "Locative media with map-based interfaces allow the user to read the city from above and to act within it, and thus 'write' it, simultaneously. These locative maps extend the reach of their users, by making complex geographical positioning data easily available at their fingertips."[32]

As technology advances, new tourist practices using locative media in urban settings are emerging. Computer-generated information extends the 'real' environment by incorporating virtual elements.[33] So-called 'augmented reality' combines virtual elements with the physical world. Information technologies and new media are densely woven into tourist practices. "[T]ourist places are now more than physical locations. They are networked and hybrid environments that can be encountered simultaneously through bodies-in-places and via mediating technologies like smartphones and social media platforms."[34] These innovative applications provide new opportunities to understand and appropriate the physical world.[35]

Self-guided tours show parallels to city tours, which have developed quite a bit and are available in many different forms today (for example, bus tours, walking tours with a local inhabitant, or tours with specific topics like street art or local cuisine). These diverse tours are popular among tourists as obtaining knowledge about the city and visiting certain parts of it represent an important part of the tourist's appropriation of an unfamiliar place. An indication of the popularity of city tours can be found on the Airbnb internet platform: besides renting apartments, this online marketplace acts as an intermediary offering city tours for interested guests. Providing information about a particular location's culture and history has always been a central part of the work of tourist service providers. However, with today's digital culture, self-guided tours offer the opportunity to explore the attractions of a city on one's own.

The following case study demonstrates a specific self-guided tour provided by Freiburg Living History.[36] Exploring the interactions between traveling bodies and the tour itself from a situated perspective, I analyze how different technological features shape the way bodies travel through and experience the historic city center of Freiburg. Freiburg Living History mainly offers guided city tours realized by professional actors who reenact scenes from medieval everyday life. A self-guided tour was added to their portfolio of city tours in 2017. This tour was developed in collaboration with Future History, a separate company that invented an interactive digital application, which would allow tourists to see how cityscapes changed by juxtaposing historical photographs with the present sites.[37] After downloading the self-guided tour to a smartphone or tablet, tourists can experience the history of Freiburg along 23 different sites in the medieval city center. Photographs, audio files with scripted reports from the perspective of historical characters, and textual descriptions foster the feeling of time travel.

14.3.1   Methods

The results presented in this chapter were obtained using a multi-method qualitative research approach. As participatory observation is the main method, the results are highly influenced by the researcher's own experiences. The preliminary guidelines for research in tourism studies are described by Tomas Pernecky and Tazim Jamal as follows: "The researcher seeks to interpret and understand the lived experience; searches for meaning, analyses, critiques, and negotiates between theory and data."[38] For this purpose, I explored the city on my own using the above-mentioned self-guided tour. I downloaded the app and followed the walking tour through the city; I took photographs, and afterwards wrote fieldnotes and detailed descriptions of my experiences and impressions. These methods, borrowed from anthropological fieldwork, resulted in a subjective evaluation combined with critical self-reflection and analysis of my actions.[39] To start from individual experiences is an appropriate method for place-phenomenology: one's own traveling body interprets sensory and social perceptions. "Place-phenomenology demands that we seek to discover the world as it is experienced by those who are involved and situated in it."[40]

Molz describes the "mobile method" of walking through a city with a self-guided tour as follows: "This doubly mobile method – involving both embodied movement around the city and virtual movement via the mobile device [...] can best be understood as both capturing *and* creating the hybrid geographies and connections with place that [the researcher] hope[s] to investigate."[41] The verbalized experiences and perceptions of the author are presented here, in addition to photos, maps, and transcripts of the audio files that are part of the Freiburg Living History application.

Additionally, I conducted in-depth interviews with the two companies that collaborated in the realization of the app. Their explanations were helpful for understanding the technological features, on the one hand, and their intentions in respect of tourists' perceptions, on the other hand. The latter were compared with the author's individual experiences on the walking tour.

14.3.2 Findings

There are three main technological features in respect of using the app and the way it helps tourists to perceive specific sites in the city: locative media such as maps, audio files with storytelling projects, and 'augmented reality' applications. In the following, these features and their impact on embodiment are presented. Farman's theoretical conception is applied in order to illustrate the embodied act with a self-guided tour.

14.3.2.1   Digital Maps – Self-Centered Wayfinding in Urban Spaces

Freiburg Living History describes the self-guided tour to potential participants as follows:

> Different characters from Freiburg's rich history will accompany you on this entertaining stroll through the old part of town. They will give you an overview of the most important events and sites of medieval Freiburg. And the best part: only you decide when you do this tour and when it's time to stop for a coffee and slice of Black Forest Cake.[42]

When the application is opened, 23 'important sites' (such as the city gates, Freiburg cathedral, or the Freiburg 'Bächle') and a suggested route are shown in the map section (see Figure 14.1). The app provides location-based services by incorporating data from 'Google Maps'. The stops are numbered, and the app suggests starting at the town hall square. However, they do not have to be visited in consecutive order, as the brief description in the app explains. The map shows the connection between the cultural sights by a pink line and one's own position via GPS. Highlighting the stops of the tour can be understood as an act of defining the site. Selecting one stop always means excluding others that may be equally worth seeing. Moreover, the user is embedded in 'blended urban geographies' which are visible on the digital interface. The map pivots around the user's position; the traveling body always marks the center point.[43] The pink line links the 23 stops (blue circles) with one another. These marked attractions can be understood from a user's perspective as 'meaningful'.

   Two main phases in using the self-guided tour can be distinguished: following the pink line means finding the next stop by walking through

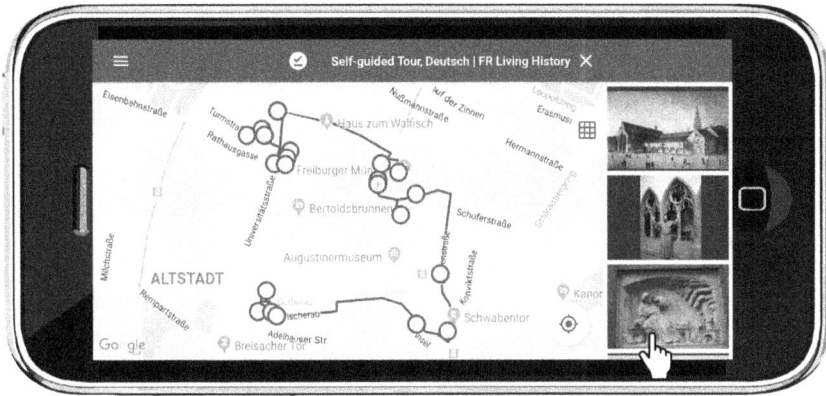

Figure 14.1 Digital map of historical sites in Freiburg (Image taken from: Rolf Mathis, Future History App. "Self-guided Tour, English. Freiburg Living History", *Rolf Mathis, vers. 10.2. Apple App Store, www .apps.apple.com/de/app/future-history/id1034236058 [12 September 2019]. Images reproduced with kind permission by Rolf Mathis).

the streets (dynamic phase), while the stories relating to the various attractions in the city's historic streets are discovered by listening to the technical device (stationary phase).[44] Maps foster the sense of where the traveling body is in relationship to its surroundings ('situatedness') and also provide a sense of direction in a particular place.[45] The map enhances the impression that the user's own location functions as the departure point for approaching any particular cultural sight during the dynamic phases.

Moreover, the top-down perspective appears as an objective representation of relational space. The impression of a relational space is intensified as the digital application provides a distance measurement from one's own position to the next site in the historic center of Freiburg. Maps, as a way of representing space, influence the way tourists move through the city. They generate new spatial configurations: "Our ideas how spaces are organized are mediated by technologies and the representations they produce."[46] The symbols shown on the map indicate tourist attractions and tram stops in Freiburg. These signs, in addition to street names, function as landmarks and help one to stay oriented. This orientation starts from our own body and relates to "something different from ourselves",[47] whether it be physical material or other bodies. In this app, the depicted landmarks provide guidance and influence our perception of space: the relational sense of direction, on the one hand, and the pre-selection of 'meaningful' sites, on the other hand, shape the self-guided tour from a traveling body's perspective. The sites highlighted on the screen (for example, historical buildings, signs with street names, tram stops) attract the user's attention; the traveling body looks for them in the urban space and turns toward them. Furthermore, vision is an active practice that goes along with movement: as we move, we actively explore the world from different perspectives. "Perceivers have an implicit, practical understanding of the way movements produce changes in sensory simulations. They also have an implicit practical understanding that they are coupled to this world in such a way that movements produce sensory change."[48] The maps and their design function as a starting point for traveling bodies to become oriented in space and encounter specific sites. The interplay between the map as a cultural artifact and the map user as a culturally inscribed body redefines the concept of territory: it becomes a relational space between the urban landscape (actual and represented) and its human user. "Representations not only frame our thinking about a space but also serve as modes of spatial production that make the space fit with our understanding of the world."[49]

Participants continuously compare the digital interface (top-down map of Freiburg's inner city) with their actual surroundings (eye-level perspective of the traveling body). Frequently, this synthesis is unquestioned and happens automatically since many people in our pervasive computing culture have learned to cope with digital maps as locative media.

> [W]e are living in a time in which the two realms of the realized and the realizing (or the actual and the virtual) do not signify themselves as exclusive spaces; instead, the interaction between these spaces continues to become mutually constructive. The reason for this transition is that the media specificity of mobile devices embodies us in a very particular way in space. The very practice of embodied space is becoming entirely reliant on the seamless interaction between our devices and our landscapes. The representation of space is not outside of the lived experience of that space. It is instead entirely incorporated into the production of embodied space.[50]

14.3.2.2 *'Augmented Reality' – Perceiving Historical Views in Urban Spaces*

Another kind of embodied engagement with the urban space is supported by an integrated augmented reality application. Users can control a slider and lay historical photographs over the actual environment they are facing. Guided by the map, the participants have to position themselves in a particular place. Interestingly, and in contrast to the top-down perspective of the general map, the activated view changes to an embodied point of view (eye-level perspective). Based on the location and the direction in which the participants are facing, they are able to overlay their current surroundings with images from the past.[51] As a result, the photograph corresponds exactly to what users see from their personal point of view. In other words, a virtual layer is blended onto the actual physical material environment. Consequently, the digital interface serves as a frame for time travel and makes the invisible visible. The digital interface or the historical image shown on the display is not isolated from the concrete and actual building. Through the digital user interface, the traveling body actively connects these two realms by visually consuming and comparing the two views. Again, this can be understood as an imaginative capacity for synthesis as digital media create a sensory experience of layering.[52]

For illustrative purposes, the medieval city gate Martinstor can serve as an example to demonstrate the functionality of the augmented reality application (see Figure 14.2). The oldest preserved city gate of Freiburg, first mentioned in records in 1238 ("Martinstor"), continues to be one of the main entrances to the city center.[53] The left frame shows the appearance of

Figure 14.2 Historical (1880) and current (2016) views of the Martinstor. By moving the slider to the left or the right, views of the same building at different times appear (Historical image taken from: C. Clare, "Martinstor um 1800", Yi Thomann and Ralf Thomann, *Alt-Freiburg in Bildern*, www.alt-freiburg.de/impressum.htm [1 October 2022]. Current image taken from: Valeria Shopina, "Martinstor 2016", *Rolf Mathis, *Freiburg Zeitreise*, Apple App Store, www.apps.apple.com/de/app/freiburg-zeitreise/id1467880678 [12 September 2019]. Images reproduced with kind permission by Rolf Mathis).

the building around 1880, while the right frame shows its modern appearance. Users can swipe to the right and a layer showing the appearance of the older building fades in. These fade-ins and fade-outs allow visual comparison of the different sizes and appearances of the building in different epochs. The same image sections are depicted in order to generate a 1:1 comparison and to give an impression of the size ratio between the historic building and the present-day building. By focusing on the battlements, one detail of the architecture, the expansion of the tower can be traced. Suddenly, a building one would perhaps pass by without a second thought receives attention – it becomes part of one's cognitive awareness. Such an experience is completely different from contrasting the two views on a computer or a laptop at home. Perceiving the changed sizes and different appearances in situ, in relation to one's own body, shapes the experience. These new ways of visualizing information via mobile technologies imbue space with additional meaning.[54] The two visualizations (the real building and the historical photograph on the device) create an impression of cityscape changes and a feeling of temporality.

14.3.2.3 *Audio Files – Experiencing Medieval Times in Urban Spaces*

A further important technological feature of the application is the opportunity to listen to audio files. Various historical characters relate their biographies and add information about buildings and former living conditions. Consequently, another layer is added to the actual physical infrastructure by stimulating the listener's imagination. The stories are told in relation to sites with which historical characters have a connection (birthplace, workplace, places where major changes in their life took place). The listeners are addressed directly by the characters ('you'), generating immediacy and involvement. The instructions one listens to serve to guide one's auditory perception. Again, the significance of site specificity should be mentioned: the tour connects the user to an "augmented city, revealing stories that would otherwise be unavailable to an unequipped pedestrian".[55] The invisible stories and descriptions of personal circumstances become part of the realm of cognitive attention and activate the user's imagination.

One primary observation is that the auditory content of locative stories affects the bodily experience of space and vice versa. While plunging into history, architecture, and culture, one is able to note the differences in today's surroundings. A strong feeling for temporalities, epochs, and societal changes emerges. "The act of storytelling is [...] an act of inscription. It is a writing of place, of identities, of relationships. It is also true that the body's relationship to specific spaces is informed out of a relationship to the narratives told about those spaces."[56]

For example, at stop number 4 of the self-guided tour, a Magister speaks about education in medieval Freiburg.[57] He explains the rules for his students and punishments for breaking them. Furthermore, the Magister describes some architectural features of Freiburg:

> The cobblestones [on the ground] are from the river Rhine. When *you* are looking around you will be able to find a lot of different mosaics. They show various motifs and tell you which guild, store or noble family is located in the building behind. This way, everybody can find their way around Freiburg, even if you cannot read or write.[58]

This short extract illustrates the function of audio files: they provide information about Freiburg's construction in an entertaining manner and help the listener to see specific features of the city. In another example, the user's attention is drawn to the mosaics on the ground that can be visually perceived so that one's imagination is activated ("What did the mosaic mean for the contemporary people?"). The traveling body then engages in a search for clues in order to read the pictorial meaning with all his or her senses. During the tour, he or she will keep an eye open for the mosaics and see them in the light of the received information. The acquired cultural knowledge thus supports the process of reading signs in the foreign city.

At the last stop, the above-mentioned Martinstor, the story of Catharina Stadellmenin, the Witch of Freiburg, is told. According to the legend, Catharina was accused of witchcraft and was hung. The audio file brings her back to life – the moment of execution and her last walk through the gate is imminent. In a haunting voice, Catharina speaks to the listener. The following is an excerpt from the audio file:

> Four things I have confessed while being tortured: First, the pact with the devil. I have formed a union with the prince of darkness. [...] Second, the devil's concubine. I consummated this union by fornication with the horned beast. Third, the bewitchment. I have vowed to bring my fellow people harm by bewitching cattle, curing children or ruining their harvest. Fourth, the Witches' Sabbath. I have convened with others of my kind at the dancing circle. [...] While the crowd screams "bloody murder" they will bring me here to this square. And execution is always a good reason to celebrate. Here the scaffold is waiting for me. I have to kneel in front of a pile of sand. The hangman's helpmates will cut my hair in the back of my neck. And then the hangman himself will show off. He's going to say: "The magistrate wants to show you mercy and give you the sword." Then he is going to bend down to whisper a blessing or an apology into my ear. And then, with one mighty swing he will decapitate me.[59]

The site appears in a new light: Catharina's story, told from her perspective, is now connected with the Martinstor as a historically portentous and emotionally moving site. A visual consumption of the building takes place, and – more importantly – it is from now on loaded with a concrete historical narrative. The historical meaning becomes tangible for the traveling body. The actor's terrified voice in combination with the scenic narration creates an oppressive atmosphere that elicits a bodily reaction. The ability to put oneself in the medieval situation increases from stop to stop of the self-guided tour. It reaches its climax at the very last stop when Catharina lets the listener experience the last minutes of her life. Strong compassion for Catharina is felt at this point of the city tour.

All in all, it is the immersive power of storytelling that leads to an experience of time travel; simultaneously there is a disconnection from the everyday life of the local residents.[60] The different epochs, life circumstances, and city usages form a sharp contrast to present-day Freiburg.

14.4   Conclusion

Tourists as traveling bodies explore urban cityscapes through different practices – primarily perception and reading processes. The body as the medium of travel encounters settings and surroundings. Various tourist services, for example, digital applications like self-guided tours, are created to support a connection between urban spaces and travelers. Practices and interactions enabled by such apps establish new and intensive relationships between human and non-human components.

The chapter has shown that maps, augmented reality, and audio files offered by a digital application can guide our perception and experience. In particular, audio files activate our imagination and show an immersive power. These storytelling projects give insight into the circumstances of medieval life, which can be experienced through the eyes of different characters. All the features generate a feeling for the atmosphere of Freiburg during the Middle Ages and simultaneously for the current situation within the city.

The multimedia presentations used in self-guided tours are only one example of innovation within the tourism industry. The latest developments in the market use advanced virtual reality technologies. The company TimeRide, for example, has invented an indoor sensory tourist experience: at the event location in Dresden, visitors climb onto a historic coach and put on virtual reality glasses that allow them to immerse themselves in the opulence of the Baroque era.[61] A ride on the coach (haptic), the 360-degree views (visual), the supporting sounds (auditory dimension), and finally participation at Fredrick Augustus I of Saxony's wedding (narrative dimension) are produced with high-tech features that evoke bodily time travel.

While TimeRide builds on individual perceptions, there is another trend toward interactive experiences. Thematic urban games (so-called Outdoor Escape Rooms) are supported by technical devices like tablets and allow users to discover cities as a group by playful means. Such games combine elements of city quiz rallies, geocaching, augmented reality technology, and further equipment, enabling the users to carry out a given mission as a team.[62] The users learn about the city's history and culture interactively, with strong bodily engagement.

The given examples illustrate current developments and show the relevance of digital applications in the tourism context. Furthermore, they illustrate the need for continued research within the dynamic field of urban tourism. This case study conducted in Freiburg shows that phenomenological approaches within human geography have the potential to understand tourist activities as embodied practices. Nevertheless, it is a scholarly desideratum to further develop both theoretical concepts and empirical methods for analyzing tourist experiences on a broader scale and to take into account different perspectives of embodied subjects and visited places.

Notes

1 Jennie Germann Molz, *Travel Connections. Tourism, Technology and Togetherness in a Mobile World* (London: Routledge, 2012), p. 43.
2 Pierre J. Benckendorff, Pauline J. Sheldon, and Daniel R. Fesenmaier, *Tourism Information Technology*, 2nd ed. (Wallingford: CABI, 2014), p. 157.
3 Stuart Eve, "Augmenting Phenomenology: Using Augmented Reality to Aid Archaeological Phenomenology in the Landscape", *Journal of Archaeological Method and Theory*, 19, no. 4 (2012): 582–600 (at pp. 586–88).
4 Jason Farman, *Mobile Interface Theory. Embodied Space and Locative Media* (New York: Routledge, 2012), p. 13.
5 I would like to thank Volkswagen Foundation for their funding of my research activities in the graduate research group "New Travel – New Media" at the University of Freiburg, Germany.
6 John Urry, *The Tourist Gaze. Leisure and Travel in Contemporary Societies* (London: Sage, 1990).
7 Tim Edensor, "Performing Tourism, Staging Tourism. (Re)producing Tourist Space and Practice", *Tourist Studies*, 1, no. 1 (2001): 59–81; and Tim Edensor, "Mundane Mobilities, Performances and Space of Tourism", *Social & Cultural Geography*, 8, no. 2 (2007): 199–215.
8 Yi-Fu Tuan, *Space and Place. The Perspective of Experience* (Minneapolis: University of Minnesota Press, 1977); Anne Buttimer and David Seamon, *The Human Experience of Space and Place* (New York: St. Martin's Press, 1980); John Pickles, *Phenomenology, Science and Geography* (Cambridge: Cambridge University Press, 1985); Yi-Fu Tuan, *The Good Life* (Madison: University of Wisconsin Press, 1986); and Nigel Thrift, *Spatial Formations* (London: Sage, 1996).
9 Kirsten Simonsen, "Bodies, Sensations, Space and Time: The Contribution from Henri Lefebvre", *Geografiska Annaler*, 87, no. 1 (2005): 1–14 (at p. 12).

10 Pau Obrador Pons, "Being-on-holiday. Tourist Dwelling, Bodies and Place", *Tourist Studies*, 3, no. 1 (2003): 47–66; Simonsen, "Bodies, Sensation, Space and Time"; Yvette Reisinger and Carol J. Steiner, "Reconceptualizing Object Authenticity", *Annals of Tourism Research*, 33, no. 1 (2006): 65–86; Kirsten Simonsen, "Practice, Spatiality and Embodied Emotions: An Outline of a Geography of Practice", *Human Affairs*, 17, no. 2 (2007): 168–81; Derek P. McCormack, "Geographies for Moving Bodies: Thinking, Dancing, Spaces", *Geography Compass*, 2, no. 6 (2008): 1822–38; Tomas Pernecky and Tazim Jamal, "(Hermeneutic) Phenomenology in Tourism Studies", *Annals of Tourism Research*, 37, no. 4 (2010): 1055–75; Kirsten Simonsen, "In Quest of a New Humanism: Embodiment, Experience and Phenomenology as Critical Geography", *Progress in Human Geography*, 37, no. 1 (2012): 10–26; Jürgen Hasse, *Was Räume mit uns machen – und wir mit ihnen. Kritische Phänomenologie des Raums* (Freiburg, München: Karl Aber, 2014); and Thomas Dörfler and Eberhard Rothfuß, "Lebenswelt, Leiblichkeit und Resonanz: Eine raumphänomenologisch-rekonstruktive Perspektive auf Geographien der Alltäglichkeit", *Geographica Helvetica*, 73, no. 1 (2018): 95–107.
11 Solène Prince, "Dwelling and Tourism: Embracing the Non-representational in the Tourist Landscape", *Landscape Research*, 44, no. 6 (2019): 731–42 (at p. 734).
12 Pons, "Being-on-holiday", p. 52.
13 Martin Heidegger, *Sein und Zeit* [1927] (Berlin: De Gruyter, 2006), p. 52.
14 Thomas Dörfler and Eberhard Rothfuß, "Place, Life-World and the *Leib*: A Reconstructive Perspective on Spatial Experiences for Human Geography", in Bruce B. Janz, ed., *Place, Space and Hermeneutics* (Cham: Springer, 2017), pp. 413–25 (at p. 416).
15 Dörfler and Rothfuß, "Place, Life-World and the *Leib*", p. 414.
16 David Seamon, "Merleau-Ponty, "Lived Body, and Place: Toward a Phenomenology of Human Situatedness", in Thomas Hünefeldt and Annika Schlitte, eds., *Situatedness and Place. Multidisciplinary Perspectives on the Spatio-temporal Contingency of Human Life* (Cham: Springer, 2018), pp. 41–66 (at p. 44).
17 Alva Noë, *Action in Perception* (Cambridge, MA: MIT Press, 2004), p. 33.
18 Jason Farman, *Mobile Interface Theory. Embodied Space and Locative Media* (New York: Routledge, 2012), p. 13.
19 Setha M. Low, "Embodied Space(s). Anthropological Theories of Body, Space, and Culture", *space & culture*, 6, no. 1 (2003): 9–18 (at p. 10).
20 Shaun Gallagher, "Interpretations of Embodied Cognition", in Wolfgang Tschacher and Claudia Bergomi, eds., *The Implications of Embodiment: Cognition and Communication* (Exeter: Imprint Academic, 2011), pp. 59–71.
21 Thomas J. Csordas, "Introduction: The Body as a Representation and Being-in-the-world", in Thomas J. Csordas, ed., *Embodiment and Experience. The Existential Ground of Culture and Self* (Cambridge: Cambridge University Press, 1994), pp. 1–24 (at p. 12).
22 Ted Toadvine, "Maurice Merleau-Ponty", *Stanford Encyclopedia of Philosophy* (2016), https://plato.standford.edu/entries/merleau-ponty/ [4 June 2022].
23 Farman, *Mobile Interface Theory*, p. 13.
24 Maurice Merleau-Ponty, *Phenomenology of Perception*, trans. Donald A. Landes (London: Routledge, 2013).
25 Jacques Derrida, *Of Grammatology*, trans. Gayatri Chakravorty Spivak (Baltimore: Johns Hopkins University Press, 1976).
26 Farman, *Mobile Interface Theory*, p. 31.

27 Jason Farman, "Stories, Spaces, and Bodies: The Production of Embodied Space Through Mobile Media Storytelling", *Communication Research and Practice*, 1, no. 2 (2015): 101–16 (at pp. 110–12).

28 Pons, "Being-on-holiday", p. 53.

29 Farman, *Mobile Interface Theory*, p. 23.

30 Noë, *Action in Perception*, p. 32.

31 Teri Rueb, "This is (not) a map", in Regine Buschauer and Katharine S. Willis, eds., *Locative Media. Multidisciplinary Perspectives on Media and Locality* (Bielefeld: transcript, 2013), pp. 137–50 (at p. 138).

32 Jen Southern, "Comobile Perspectives", in Buschauer and Willis, eds., *Locative Media*, pp. 221–42 (at p. 222).

33 Eve, "Augmenting Phenomenology", p. 587.

34 Molz, *Travel Connections*, p. 42.

35 Johanna Brewer and Paul Dourish, "Storied Spaces: Cultural Accounts of Mobility, Technology, and Environmental Knowing", *Human-Computer-Studies*, 66, no. 12 (2008): 963–76 (at p. 996).

36 Freiburg Living History, "Self-guided Tour" (2019), www.freiburg-living-history.de/self-guided-tour/ [8 September 2019].

37 Future History provides an app infrastructure and technology that is used by different partners in the tourism industry. The partners (for example, destination management organizations or companies offering city tours) use the software and create the contents for tours themselves in order to present single sites or a whole city district to visitors. For German cities like Leipzig, Magdeburg, and Konstanz, digital versions are already available in several languages.

38 Tomas Pernecky and Tazim Jamal, "(Hermeneutic) Phenomenology", p. 1067.

39 Clifford Geertz, *The Interpretation of Cultures. Selected Essays* (New York: Basic Books, 1973).

40 Dörfler and Rothfuß, "Place, Life-World and the *Leib*", p. 423.

41 Molz, *Travel Connections*, p. 27.

42 Rolf Mathis, Future History App. "Introduction. Freiburg Living History", *Rolf Mathis, vers. 10.2. Apple App Store, www.apps.apple.com/de/app/future -history/id1034236058 [12 September 2019].

43 Jason Farman, "Map Interfaces and the Production of Locative Media Space", in Rowan Wilken and Gerard Goggin, eds., *Locative Media* (London: Routledge, 2014), pp. 83–93 (at p. 88).

44 Anja Stukenbrock and Karin Birkner, "Multimodale Ressourcen für Stadtführungen", in Marcella Costa and Bernd Müller-Jacquier, eds., *Deutschland als fremde Kultur: Vermittlungsverfahren in Touristenführungen* (München: iudicium, 2010), pp. 214–43 (at p. 220).

45 Farman, *Mobile Interface Theory*, p. 45.

46 Brewer and Dourish, "Storied Spaces", p. 965.

47 Simonsen, "In Quest of a New Humanism", p. 19.

48 Noë, *Action in Perception,* p. 66.

49 Farman, "Map Interfaces", p. 85.

50 Farman, *Mobile Interface Theory*, p. 46.

51 Farman, *Mobile Interface Theory*, p. 40.

52 Farman, "Stories, Spaces, and Bodies", p. 107.

53 Future History, "Martinstor" (2019), www.future-history.eu/de/ansicht/martin-stor-freiburg-im-breisgau-1880-Joescheck [4 June 2022].

54 Farman, *Mobile Interface Theory*, p. 40.

55 Molz, *Travel Connections*, p. 49.

56 Farman, *Mobile Interface Theory*, p. 118.
57 Mathis, "Magister. Freiburg Living History" [12 September 2019].
58 Mathis, "Magister" [12 September 2019], 00:01:22-55.
59 Mathis, "The Witch of Freiburg. Freiburg Living History" [12 September 2019], 00:00:33-01:57.
60 Molz, *Travel Connections*, p. 48.
61 TimeRide, "TimeRide – Mitten im Damals" (2019), www.timeride.de/en/ [4 June 2022].
62 ENMAZE, "ENMAZE Escape Room Heidelberg" (2019), www.enmaze-hei-delberg.de [4 June 2022].

Index

For Product Safety Concerns and Information please contact our EU
representative GPSR@taylorandfrancis.com
Taylor & Francis Verlag GmbH, Kaufingerstraße 24, 80331 München, Germany